Life Out of Bounds

LIFE OUT OF BOUNDS

Bioinvasion in a Borderless World

Chris Bright

The Worldwatch Environmental Alert Series
Linda Starke, Series Editor

W. W. NORTON & COMPANY
New York   London

Worldwatch Database Disk

The data from all graphs and tables contained in this book, as well as from those in all other Worldwatch publications of the past two years, are available on disk for use with IBM-compatible or Macintosh computers. This includes data from the State of the World *and* Vital Signs *series of books,* Worldwatch Papers, World Watch *magazine, and the* Environmental Alert *series of books. The data are formatted for use with spreadsheet software compatible with Lotus 1-2-3 version 2, including all Lotus spreadsheets, Quattro Pro, Excel, SuperCalc, and many others. To order, send check or money order for $89 plus $4 shipping and handling, or credit card number and expiration date (Mastercard, Visa, or American Express), to Worldwatch Institute, 1776 Massachusetts Ave., NW, Washington DC 20036. Or you may call 1-800-555-2028, fax us at 1-202-296-7365, or reach us by e-mail at wwpub@worldwatch.org.*

Contents

Acknowledgments

In July 1995, Anne Platt McGinn and I hosted a Worldwatch press conference on bioinvasion. The occasion for the briefing was the release of an issue of the Institute's magazine, *World Watch,* which carried articles by both of us. Anne's was on the ecology of infectious diseases—on how environmental degradation tends to spread infection. Mine was on the spread of exotic species—on the damage that can occur when a species of mosquito, say, or a weed moves into new territory.

In the course of writing those articles, it had become apparent to both of us that we were really talking about the same phenomenon: about how human activity is "stirring up" the Earth's organisms—whether viruses, weeds, or whatever—and about why the consequent levels of biotic mixing tend to injure our societies and the natural world. The press conference was our way of inviting the public to explore this terrain with us. *Life Out of Bounds* is a continuation of that process.

This book began as a form of collaboration and as it grew, it seems to have developed an ever wider cast of contributors. I have never written a book before, and I had no idea how many debts I would accumulate before my pages saw the light of day. I owe, first, an enormous intellectual debt to a number of writers and researchers I have never met. Among them is Alfred Crosby, whose *Biological Imperialism* showed me how profoundly "political" so many apparently natural landscapes really are. Edward Tenner's *Why Things Bite Back* introduced me to the idea of "revenge effects" in both the technological and the natural world. The authors of the seminal 1993 U.S. Office of Technology Assessment report on bioinvasions helped me see the economic dimensions of the problem.

My work also rests heavily on James Carlton's extensive and fascinating coverage of marine bioinvasions, on Christopher Lever's encyclopediac surveys of introduced animals, and on the papers of many of the presenters at the Norway/UN conference on exotics held at Trondheim in 1996. In various ways and at various points, I depend on many other scientists and journalists whose works are cited in my notes. I'm also in debt to one eminent writer whom I have met: Bill McKibben, who took time out to look over a draft of the book, and whose kind encouragement was deeply appreciated.

Several experts in the field also gave me crucial advice on how to revise an earlier draft of this work. Faith Campbell's knowledge of exotic plants and the politics surrounding them has been an invaluable resource. Bruce Coblentz's thoughtful comments on terrestrial animal invasions prevented me from going astray on more than one point. I owe much to Peter Jenkins for his patient explanations of the politics, economics, and on-the-ground realities of the issue. Maureen Kuwano Hinkle's expertise in agriculture cleared a path for me through some of the thornier ground in that field. And Peter Raven's stimulating last-minute review helped me rethink several thematic concerns. I'm extremely grateful to them all.

The book bears the stamp of many of my colleagues within the Institute as well. John Tuxill provided a trenchant and thorough analysis of the entire work, and David Malin Roodman offered a valuable critique of several chapters. I'm grateful to Hilary French for a useful review of the entire text, and in particular for her comments on the final chapter. Ashley Mattoon helped me to a clearer understanding of the problems that beset industrial forestry. At several points in my argument, I lean heavily on the research of Anne Platt McGinn, my copresenter at that 1995 press conference. I profited greatly from Gary Gardner's advice on agriculture and especially on how to use agricultural research resources. And I'm grateful to a fellow editor on the *World Watch* magazine staff, Curtis Runyon, for taking time out from his cluttered schedule to read and comment on the book's language.

I'm also indebted to several colleagues who built a good deal of the "infrastructure" of this project. Lori Brown, the Institute Librarian, showed a degree of ingenuity that seemed almost magical, to me at least, when it came to tracking down what were surely some of the most bizarre documents ever requested from her. Assistant librarians Laura Malinowski and Anne Smith have been indefatigable in their efforts to sort out my often rather confused research requests and to prevent me from running up library fines that would have been little short of criminal. I'm grateful to my research interns, Candace Chandra and Sophie Chou, for their good humor, their determination to get the facts straight, and their ability to think on their feet. I've done enough fact hunting and checking of my own to know how much stamina and ingenuity the task requires.

I'm deeply grateful to the Foundation for Ecology and Development for its support of this project in the form of a grant. Investing in ideas is surely one of the toughest missions in the nonprofit sector, but it is also one of the most important. More generally, research at the Institute is supported by the following: the Nathan Cummings Foundation, the Geraldine R. Dodge Foundation, the Ford Foundation, the William and Flora Hewlett Foundation, the W. Alton Jones Foundation, the John D. and Catherine T. MacArthur Foundation, the Charles Stewart Mott Foundation, the Curtis and Edith Munson Foundation, the David and Lucille Packard Foundation, The Pew Charitable Trusts, the Rasmussen Foundation, the Rockefeller Brothers Fund, Rockefeller Financial Services, the Summit Foundation, the Surdna Foundation, the Turner Foundation, the U.N. Population Fund, the Wallace Global Foundation, the Weeden Foundation, and the Winslow Foundation. In addition, the support of numerous individuals through the Friends of Worldwatch campaign is deeply appreciated. Special thanks also go to the newly established Council of Sponsors: Tom and Cathy Crain, Toshishige Kurosawa, Kazuhiko Nishi, Roger and Vicki Sant, Robert Wallace, and Eckart Wintzen.

Finally, I owe an enormous debt to several senior

Institute colleagues. I depend both explicitly and implicitly on the work of Janet Abramovitz, whose knowledge of global biodiversity issues has been a constant support and something I've almost come to take for granted. Ed Ayres edited the magazine article from which this book sprang and has worked with me ever since, first on a conceptual level, and then with my writing—helping me find the language that will, I hope, help this book find an audience. I could not have written the book without his moral and intellectual support.

Christopher Flavin, senior vice president for research, also played a critical role in virtually every step along the way. Chris showed me how to build my project into the Institute's interdisciplinary mission. He insisted on pushing me beyond my intellectual "comfort zone," and onto new ground. Chris's brand of genial skepticism is one of the driving strengths of our research program. Linda Starke, the editor of this series, has won my enduring respect, not only for her deft pen but for her ability to handle one of the Institute's most exasperating authors. And like everyone here, I owe a great deal to Lester Brown, for founding the Institute and for letting it grow. There are few if any other organizations at which this kind of book—at least with this kind of author—could have found a home.

Finally, before I offer my work to the public, I offer it to my son Matthew, who is teaching me continually to see the world with new eyes.

Chris Bright
Worldwatch Institute
1776 Massachusetts Ave., N.W.
Washington, D.C. 20036

June 1998

Foreword

Globalization is much in the news these days. Everything from cars to tennis shoes is manufactured thousands of miles from customers, and money moves instantaneously from a bank account in India to one in Switzerland. The ramifications of this brave new global world are understandably causing some people concern, especially after the collapse of the Asian financial markets in late 1997 revealed the vulnerability of the world financial system.

But too few people have noticed another, perhaps more frightening form of globalization: the movement of exotic plants and animals into virtually every ecosystem on Earth. The ability of pests, weeds, and dangerous pathogens to move around the world today is truly staggering. Brown tree snakes hitchhike from Guam to

Hawaii hidden in the wheel wells of a jet. Zebra mussels get swept up in the ballast water of a supertanker and find a new home, and new victims, in the Great Lakes. The Asian tiger mosquito, a major carrier of dengue fever, encephalitis, and yellow fever, moves from country to country with ease in containers of used tires.

Our vulnerability in this case is not just a matter of shaky financial markets. It affects the very underpinnings of all economic activity—the stability and integrity of the Earth's land, forests, and waters, and all that we reap from them. In *Life Out of Bounds*, Chris Bright describes these escalating threats to the foundations of biological diversity and productivity, their sources, and why they should be of concern to us all. And he lays out the steps that need to be taken to address the problem of biological globalization—from international codes of conduct right down to being aware of what is in our own backyards.

This tenth volume in the Worldwatch Environmental Alert Series follows David Roodman's *The Natural Wealth of Nations*, which explored another long-ignored policy issue: how society's fiscal policies affect the health of our basic life-support systems. (See page 2 for a list of all the titles in the series.) We hope that our contributions on these topics will help foster more detailed public discussion of how to build a sustainable society. We welcome your reactions.

Linda Starke, Series Editor

Life Out of Bounds

1

Evolution in Reverse

Some 240 million years ago, well before the reign of the dinosaurs, all of the Earth's major landmasses were locked into a single continent. A monstrous plaque of rock called Pangaea sat alone amidst the waters of an even more monstrous planetary ocean. Eventually, Pangaea fragmented. At geologic pace, its shards sailed out over the blank blue immensity to create, for the human moment, the present continental configuration.[1]

The macro structure of the planet might seem to be the one aspect of the world that people cannot change. And yet the currents of human movement are beginning to alter the ancient evolutionary function of the planetary surface. As vessels for the natural communities that evolved within them, the Earth's pockets and humps, its wet and dry places, are losing their integrity—their separateness. At a frenetic and ever-increasing pace, the global economy is merging the world's ecosystems, smearing them into each other. We are in the throes of a vast and little-noticed biotic upheaval. As ecological entities, the continents are coming together again; the seas are spilling into each other. And this biotic turmoil is reaching levels of disturbance that no actual meeting of rock or water could possibly achieve. Modern commerce is wrapping the world's natural systems in a web of connections that is far more comprehensive than anything that could have existed on the ancient super-continent. A kind of hyper-Pangaea is emerging.

The physical roughness of the Earth—its structural variety—has tended to hold its living communities in place. The barriers that surround any particular ecosystem help set the terms of life within it. They tie a particular assemblage of plants and animals together, and they tend to exclude predators, competitors, and diseases that evolved elsewhere. Islands provide the extreme case. In their isolation, many island creatures have evolved into forms found nowhere else—the giant tortoises of the Galápagos, for example, or the colorful "picture-winged" fruit flies of Hawaii.

The planet is scored by thousands of more subtle barriers too. A "rain shadow" downwind from a mountain ridge may be too dry for forest; an ocean current may isolate two distinctive coral reefs. Even the lives of highly mobile creatures are likely to be governed by barriers of one sort or another. The salmon that hatch in the rivers of western North America may swim together in the ocean, yet each strain returns to its own river to breed, thereby preserving its distinctive genetic identity. And so from one horizon to the next, a subtle matrix of barriers has allowed the communities of life to work out evolutionary answers to a particular spot of land, a stream, or a set of ocean currents. Natural barriers are the instruments of evolution.

Today these barriers are losing their ecological reality, as more and more organisms are moved around them. A western Atlantic jellyfish, for example, is pumped out of a ship's ballast tank and into the Black Sea, where it wrecks the fisheries. Escaped garden plants strangle North American wetlands and rare island forests. Plantations of Australian eucalyptus trees displace native forests throughout the developing world—and sometimes native forest peoples as well. The farming of commercial shrimp species obliterates coastal fisheries and the local economies that depend on them. Range-devouring weeds sprout from contaminated crop seed; virus-laden mosquitoes emerge from shipping containers. In these and hundreds of other ways, the silent uproar of biotic mixing is damaging both worlds that people inhabit: the natural and the social.[2]

Some degree of movement through the Earth's barriers has always been a part of life, of course—no natural community is hermetically sealed. A shift in the prevailing sea breeze might bring a colony of bats to an island; an increase in rainfall might allow forest out onto a prairie. But the artificial Pangaea that we are creating differs fundamentally from such natural range extensions in three respects:

- *In the frequency of movement.* Under natural conditions, the arrival of a new organism—an "exotic species"—was in most areas a rare event. Today, it can happen any time a ship comes into port or an airplane lands. In places where the current rate of arrival has been estimated, it generally appears to be thousands of times faster than the previous natural rate.

- *In the pervasiveness of movement.* In the past, a major ecological change sometimes allowed a mixing of one biota (the set of plants and animals native to an area) with another. One of the most dramatic episodes involved Beringia, the ancient land bridge that once linked Siberia to Alaska. Over the course of many millennia, Beringia admitted numerous Eurasian species (including people) into the New World. Today, intense biotic mixing has moved from being an occasional regional event to a chronic global phenomenon.[3]
- *In the fact that "impossible migration" is now not only possible, but common.* Under natural conditions, the planet's physical structure imposes formidable barriers to certain types of movement. Bounded by 6,000 kilometers of salt water, for example, or 1,000 kilometers of desert, many organisms would live out their evolutionary lives without crossing to terra incognita on the other side. Today, such crossings are routine. Water hyacinth, an aquatic weed that is suffocating East Africa's Lake Victoria, comes from South America; the disease that is killing off the crayfish in European streams comes from the crayfish that live in North American streams; melaleuca, a tree that is invading the Florida Everglades, is from northern Australia.

The effective collapse of the world's ecological barriers is a phenomenon, so far as we know, without precedent in the entire history of life. (The real Pangaea, of course, would have had plenty of very durable barriers.) During the past several centuries, and today at an ever-increasing rate, the Earth's natural communities are being disrupted by exotic species—organisms that have crossed those barriers to take up residence in ecosystems where they did not evolve. Bioinvasion, the spread of exotics, is fast becoming one of the greatest threats to the Earth's biological diversity.

As a global threat of extinction, bioinvasion may already rank just behind "habitat loss"—a much more general category that can be taken to include almost any kind of physical disruption. For certain types of organisms, exotics are clearly the principal threat: during the past century in the United States, for example, exotics have been a factor in 68 percent of fish extinctions. And increasingly, these two forms of ecological decay appear to be merging into a single syndrome. As more and more habitat is burned or bulldozed away, the remnant natural areas grow ever more vulnerable to invasion. The wilds of the new millennium are melting into degraded landscapes infested by exotic weeds, weakened by exotic pathogens, chewed over by exotic browsing mammals. (See Chapter 5.)

Although the process often ends in extinction, current extinction statistics do not come close to capturing the full dimensions of the problem. That is because exotics frequently suppress large numbers of native species without pushing them completely over the brink. Take the melaleuca, the tree that is invading the Florida Everglades. Undisturbed Florida wet prairie typically contains 60–80 native plant species, while an area covered by a melaleuca thicket usually contains only 3–4, if that. Successful invasions often cause "functional extinctions" like this. The native species may still exist, but over much of the terrain they are growing at densities too low to perform their former ecological role—for instance, as food plants for native animals.[4]

Bioinvasion is perhaps the only category of environmental degradation that can corrode virtually every level of biological organization. On a broad landscape level, exotics like the melaleuca can replace entire communities of native plants and animals. At the other end of the spectrum, interbreeding between an exotic and a native relative can unleash a "genetic invasion" that undermines a native gene pool. In western North America, for example, mass releases of hatchery-bred salmon have undermined some wild salmon stocks. In such cases, the native in effect becomes exotic.[5]

The cultural effects of exotics can be as profound as the biological ones. Human pathogens, for instance, travel as readily as crop pests or weeds, and entire branches of humanity have fallen away as a result. The diseases brought into the Americas by European colonists precipitated one of the greatest cultural crises in history, and the spasms of that crisis reach right into the present. In the century following the conquistadors' arrival, as many as two thirds of the western hemisphere's native inhabitants—perhaps 30 million people—may have succumbed to smallpox, malaria, and various other Old World diseases—diseases to which they had almost no resistance. To a considerable degree, the Europeans inadvertently "created" the wilderness they then went on to explore. Today, miners and settlers continue to spread these pathogens to the native peoples of the Amazon basin, with disastrous effect. Since the mid-1980s, for example, about a quarter of the Yanomami people have succumbed to exotic diseases.[6]

The exploding volume of migration and travel is now

pulling most of humanity into a single microbial system, and no society may really be prepared for the results. Epidemic cholera has recently returned to the Americas, yellow fever may be poised to invade Asia, and we have barely begun to identify the malign synergies produced by overlapping epidemics.

Nor are the social effects of bioinvasion limited to disease. Exotics ruin our crops and suppress our fisheries. They cause declines in forest and rangeland productivity. Aggressive exotic water weeds and shellfish are fouling dams, power plant intake pipes, and irrigation canals. Some exotic plants are increasing the rate and intensity of brushfires; others are dropping water tables. In these and many other ways, exotics are costing societies all over the world billions of dollars every year.

All of this damage, both natural and cultural, results from a process that is profoundly "counterintuitive." Conceptually, the problem of invasion comes down to this: why should adding a species to an area end up *reducing* that area's biological diversity? The exotic dandelion on your lawn, for example, is presumably just one more species in the local plant community—its only known victims are people who value uniform grass. And most exotics never even make it to dandelion status. Most do not succeed in establishing themselves in their new ranges—they just die out. Even those that do establish themselves will not necessarily have a detectable ecological effect.

But the paradox disappears once you look from the individual exotic to the process as a whole. The proportion of exotics that cause serious trouble is difficult to estimate, but a very rough rule of thumb, sometimes called the "tens rule," is that 10 percent of exotics introduced into an area will succeed in establishing breeding populations, and 10 percent of those will go on to launch a major invasion. When that happens, the exotic has graduated from dandelion to melaleuca status. It has escaped the predators, diseases, and other factors that kept it in check in its native range, and has found nothing comparable in its new range. It is facing organisms that did not evolve in its presence and that may not be adapted to competing with it or escaping from it. This scenario keeps replaying itself all over the world; the result is usually lots of exotic and a lot less of everything else.[7]

Since the global economy is continually showering exotics over the Earth's surface, there is little consolation in the fact that 90 percent of these impacts are "duds" and only 1 percent of them really detonate. The bombardment is continual, and so are the detonations.

Many invaders seem to owe their explosive ecological power to a complex of traits known as "weediness." Weedy invaders mature quickly, multiply prolifically, spread easily, and often do especially well in disturbed conditions. Animal "weeds" tend to be highly adaptable in their diets. The ubiquitous rats and house sparrows, the zebra mussels invading North American waterways, the water hyacinth and the melaleuca—all these organisms are weeds.

As the weeds spread, displacing more and more local diversity, the world becomes a steadily more homogenized place. The same weeds begin to crowd every rangeland; water hyacinth smothers warm-climate waters the world over. Goats gnaw the shrubbery to stubble on island after island. In all of these places, as the local creatures disappear, the ecosystem they formed a part of tends to weaken. An artificially simplified community, like a machine that is missing a lot of its parts, is more likely to break down. A fire, say, or an outbreak of disease that might have had little effect in a healthy, complex community may seriously disturb a simplified, sick one. And that disturbance will lay the area open to yet more invasion. This is the cycle of degradation that is coming to characterize our era.[8]

★   ★   ★   ★

Bioinvasion is now a profound and global challenge to our economic system, to our technical conservation skills, and to our ethics—our ability to recognize a "right to existence" in other living things. Yet policy responses to the threat have generally been weak and uncoordinated. Only the worst invaders get serious attention, and even then there is rarely any systematic inquiry into the social and economic processes that launched the invasion in the first place.

To some degree, this lack of response can be explained by the enigmatic qualities of the problem itself. Despite 40 years of study, ecologists have not been able to discover natural "rules" that govern the processes of invasion and that have any real predictive value. Bioinvasion is a deeply unsatisfying policy topic. It is messy, frustrating, depressing, and unpredictable: it does not lend itself to neat solutions. Consider, very generally, what we don't know.[9]

We don't know *which* organisms will become successful invaders. No common characteristic has been detected. It is true that many of the worst invaders are highly adaptable "generalists"—weeds, in other words. But there are "specialist" invaders too. Some invaders have huge home ranges; some have very small ones. Some have close relatives that are also dangerous, while the closest relatives of others do not seem to be invasive at all. And some very aggressive invaders may actually be retreating in their home ranges. The melaleuca, one of South Florida's nastiest pest plants, is being crowded in its native northern Australia by catclaw mimosa, a spiny South American shrub—and by pond-apple, a native of the Florida Everglades.[10]

We don't know *where* invasions will occur. True, disturbed ecosystems are generally more vulnerable to exotics than intact ones. One of the reasons that Eurasian cheat grass now dominates 25 million hectares of western North America is almost certainly that ranchers allowed their cattle to overgraze the native grasses. But as with the "weediness" rule, there are all kinds of exceptions. In surviving tracts of undisturbed Hawaiian rainforest, for example, the dominant insects are now frequently exotic. In the Great Lakes, water quality improvements have probably helped the sea lamprey, a predatory exotic fish, since lamprey larvae are fairly sensitive to pollution.[11]

We don't know *when* an invasion will occur. Many exotics probably find their way into a new range several times before they succeed in establishing themselves. Even then, an exotic may spend decades as an innocuous good citizen in its new home before some subtle adaptation or some shift in the ecological dynamic triggers an explosive invasion. This "incubation period" is so common in plant invasions that it has a kind of reverse predictive value: there are almost certainly many more exotic plants out there than anyone has noticed. According to one expert on weed invasions in the United States, exotic weeds usually have to be in the country for 30 years or to have spread to more than 4,000 hectares before they are even discovered. (See Figure 1–1 for a graph of a typical plant invasion.)[12]

We don't know *what* an invasion will do. Because invasive exotics can do a great deal more than simply displace native species, they have a considerable capacity for surprise. Take, for example, the case of the tiny opposum shrimp in Montana's Flathead River system. Wildlife officials introduced the shrimp around 1970 to increase the forage base for the kokanee salmon, another introduced species. But salmon tend to feed near the surface and the shrimp only rose to the surface at night, when the salmon could not see them. So the salmon could not eat the shrimp, but the shrimp ate all the plankton that the salmon fry depended on. The salmon population crashed, then the bears, birds of prey, and other creatures that had come to depend on the salmon disappeared. A tiny shrimp had starved eagles out of the sky.[13]

Number of Topographic
Quadrangles Occupied

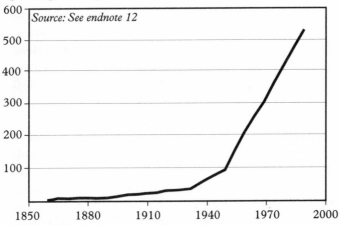

FIGURE 1–1. *Spread of European Mustard into Eastern Canada and the United States, 1860–1990*

Exotics, by and large, seem to make up their own rules. We invite them to play but they name the game. What will this creature do if it lands in that spot? About all we can say with assurance is this: if it's causing trouble somewhere, then you don't want it anywhere else. Bioinvasion may be the least predictable of all the major forms of environmental disruption.[14]

It may also be the hardest to fix. In general, the key to dealing with environmental problems—habitat destruction, say, or pollution—is to stop the offending activity. Granted, that is usually no mean feat. But if it can be achieved, natural processes, with thoughtful management, will then heal the system. Time, however, does not heal invasions. An intense invasion may "peak out" and subside after exhausting most of the local resources, but that does not mean the exotic will go away. It may rebound when its food supply recovers, or it may spread elsewhere. So while an oil spill that occurred 20 years ago is probably not a pressing concern today, there are hundreds of invasions that began more than a century ago and that are desperately urgent problems right now. This "biological pollution" is smart pollution. It adapts, it looks for ways to survive, and instead of diminishing over time, it usually entrenches itself.[15]

Beyond the tortured topics of invasion ecology, there is another reason why exotics have attracted so little in the way of policy response: they are so common, such a standard part of our environment, that their presence is not usually suggestive of dysfunction. The tendency to spread exotics is a deeply embedded and nearly universal aspect of culture. For thousands of years, people all over the world have deployed exotics for both necessity and pleasure—to feed themselves, to mold the landscapes in which they live, to stock their gardens, rivers, and forests. Accidental releases seem to be a cultural constant as well. Humanity has always been a wandering species and we have offered the globe to those creatures that profit from our presence—our diseases and parasites, our crop pests, our mosquitoes and fleas. Nearly everywhere, invasion has been, in varying degrees and in different ways, a standard feature of the human past. But as with other forms of environmental degradation, there is a big difference between regional or low-level pressure and what happens when the process gains intensity and goes global. Biotic mixing on a global level began in earnest five centuries ago, as the Age of Discovery dawned. It is reaching its logical extreme today, in the emergence of a global economy. At its current level, it is no more sustainable than are current levels of deforestation or atmospheric carbon emissions. Bioinvasion has become another way of measuring the unsustainability of the contemporary economic order.

At its current rate, bioinvasion may not be "culturally sustainable" either. The march of the weeds is robbing the landscapes in which we live of their "naturalness"—of their power to reflect something other than our own mismanagement. Invasion threatens us with a kind of culturally deadening solipsism, in which it becomes harder and harder to experience nature as distinct from ourselves. Humanity is not meant to be a patient in a sickroom, with nothing to contemplate but our own diseases. We came into being out-of-doors; our social and psychological welfare may be linked in ways we cannot fathom to the welfare of nature as a whole. We may need the "otherness" of nature just as much as we need clean water and air.

But human memory is so short compared with the scale on which nature operates. Who can remember where each weed originally came from? And it is tempting to speed the process of forgetting, by rechristening any entrenched exotic as a native. Capitulating to the invader may often seem the most realistic course of action. No doubt, nature eventually will transform many of these organisms. The scattered populations of many exotics may ultimately go their separate evolutionary ways and become distinct species, each within its own native range. (Some plant diseases may already be doing this, but that is not good news: each new version of such a pathogen is another potential invader.)

For most types of organisms, however, the process will not occur on a time scale that can matter much to those of us alive today. In the meantime, blurring the distinction between native and exotic is a tactical mistake because it invites people to view every invader as just a native in the making. Given the current levels of biotic mixing, that is a little like dismissing AIDS with the notion that one day that virus too may evolve into something more benign.

Besides, capitulation is not necessary. Despite the formidable ecological and social difficulties of countering invasions, we already have the tools necessary for rapid progress. Our principal challenge now is not so much technical as cultural. Our contemporary global reach is no longer compatible with an invasion mentality—an attitude toward nature that accepts invasion as inevitable or even desirable. The key to discarding that mentality is a kind of historical consciousness, an awareness of how the social machinery of invasion was built in the first place. Our current problems have been a long time in the making. To deal with them effectively, we will need to understand not just present ecological conditions, but the history of our "invasion cultures."

Contemporary ecology and the cultural past: all the following chapters deal with these two topics, each from a different perspective. Part I is concerned primarily with invasion as an ecological process. Chapters 2, 3, and 4 survey three very broad ecosystem categories, along with the primary human enterprise in each—prairies and agriculture, forests and forestry, the waters and fisheries. Chapter 5, "Islands," explores the type of landscape that has suffered the most from invasion, and presents it as a model for the problem as a whole. Part II looks at invasion primarily as a cultural process. It begins with two short cultural histories of invasions, first the intentional ones (chapter 6), and then the accidents (chapter 7). Chapter 8 sketches out the economic implications of the current situation and considers the global economy itself as a homogenizing force. In Part III, the book's final chapter reviews the resources at our disposal—legal, political, ecological, and personal—for treating the planet's invasion disease.

I

The Geography
of Invasion

2

The Fields

One of the most powerful bonds in the history of life is the alliance between grasses and people. During the past 10 millennia, in at least seven separate times and places, people have invented farming, and farming is essentially a managed invasion of edible plants. The farmer's most productive plants have generally been those with edible seeds, and nearly all of these are grasses, such as corn, wheat, rice, and barley. The fundamental architecture of civilizations is not to be found in buildings but in fields—in our vast artificial prairies of domesticated grasses.[1]

If you take a broad ecological view of the process, you might argue that the grasses have done a wonderful job of domesticating us. Most crop plants are utterly dependent on an enormous agricultural infrastructure that feeds and waters them, protects them from pests, and coddles their germplasm. Yet they are among the most successful organisms on Earth. Corn, for example, was first domesticated around 5,000 years ago from a grass that grows in scattered patches through the rugged mountain country of southern Mexico. Today, corn usually carpets about 140 million hectares of the Earth's surface, an area the size of Germany, France, and Spain combined. Overall, the cereal grains provide some two thirds of humanity's calorie intake (directly and through grainfed livestock) and occupy about half of the world's arable land.[2]

But the success of the grasses inevitably entails the success of their pests. As the edible prairies spilled over the continents, hordes of insects, weeds, and pathogens followed. The weeds were usually in the vanguard. At the dawn of the sixteenth century, as European societies began to expand abroad, hundreds of Eurasian weed species arrived on foreign shores as contaminants of seed grain and fodder. These weeds, toughened by some 10 millennia of contact with European cropping and grazing techniques, burst onto foreign soil and found little to stop them. Eventually, for instance, in most regions where the climate matched that of the Mediterranean (central Chile and parts of South Africa, California, and southern Australia), the major grasslands were transformed into a kind of degraded version of the Mediterranean flora. (See Figure 2–1.) Much of the temperate-zone prairie outside Eurasia has been "homogenized" in this way. Even where the crops themselves are not growing, their weeds usually are.[3]

And while the colonial era has long since receded as far as European peoples are concerned, it is still very much in force when it comes to European plants. In the

Percent

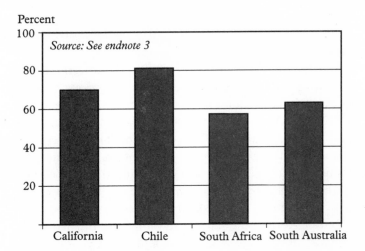

FIGURE 2–1. *Eurasian Plants as Share of Total Exotic Flora in the Mediterranean Climate Zones of Four Regions, 1990s*

western United States, for example, one of the most successful colonists has been a Eurasian weed called cheat grass. Like many highly invasive plants, cheat is fire-adapted: it spreads by burning. Fires are natural events in most landscapes, apart from the coldest and the wettest ones, but the natural rhythms of burning vary a great deal. Fire may visit a swath of prairie every year or two, but leave a northern black spruce bog undisturbed for centuries. Fire-adapted grasses like cheat usually come from landscapes with very rapid fire cycles. Fire is their native element and they take it with them when they invade. After the growing season, they cure to a fine, dry tinder, so they tend to burn every year. They recover quickly from fire, but in areas where the natural fire cycle is slower, the native vegetation generally does not. The result is a vicious circle of invasion, fire, retreat of the native plant community, then more invasion.[4]

Cheat was first noticed in 1889 in southern British Columbia. It was soon popping up in various wheat-growing areas in the arid intermountain west, and by 1930 it had burned its way throughout the sagebrush biome, which stretches from the Canadian border into Nevada and Utah. Today, it is present on more than 40 million hectares, an area about twice the size of the state of Nebraska. It dominates nearly two thirds of that, and the fires it feeds are continuing to spread it. The rich flora of the sagebrush biome has burned away to yield little more than cheat and silence. Songbirds disappear for lack of cover and forage. Deer and antelope starve. As the rabbits and rodents die off, so do the birds of prey that hunt them. Cheat and other rangeland weeds are spreading over the western United States at an estimated rate of 1,840 hectares per day.[5]

The artificial prairies attracted animal "weeds" as well. Rats and mice, for instance, attacked the crops both in the fields and in the bins, and were recognized as a bane very early. Middle Eastern farmers were using pottery rat traps as early as the third millennium B.C. During the past 2,000 years, two rodent species from southeast Asia, the black rat and brown rat, became the most successful vertebrate pests the world has ever seen. A combination of intelligence, adaptability, and reproductive prowess allowed these species to ride the tide of colonial expansion to virtually every spot it reached and often far beyond. In the process, they have appropriated a substantial share of the agricultural enterprise. Rodents probably consume about 20 percent of the world's grain harvest and perhaps as much as 75 percent in some African countries.[6]

Most of agriculture's other animal weeds are insects. Some of these creatures have ridden to global prominence as part of a crop's "native" pest burden, but others just got lucky when a crop they could eat was brought to them. When corn, for example, reached Europe, it encountered a little Eurasian moth that had until then been a relatively minor agricultural nuisance. Today, the European corn borer is a major scourge with a worldwide distribution. It claims roughly 7 percent of the world's corn and occasionally much more in particular regions. In some areas of Canada, for instance, it destroyed roughly 30 percent of the 1995 harvest.[7]

To their artificial flora, colonial farmers added their familiar artificial fauna. Cattle, horses, goats, sheep, and hogs became the dominant herbivores in agricultural landscapes all over the world. But unlike most of the major crops, the animals could generally survive on their own—and that is what they were often expected to do. Casual release of livestock has always been a part of frontier life in the New World. The seventeenth century English Puritans released their hogs into the New England forests to "breed in great numbers by reason of the abundance of acornes, [and] groundnutts," in the words of John Winthrop, governor of Massachusetts Bay colony. Essentially the same practice can be found among modern ranchers in western North America or on the fringes of Amazonia.[8]

The resulting explosion of wild livestock rivaled the weed invasions. By 1700, for instance, Virginia colonists were complaining that wild pigs "swarm like Vermaine upon the Earth." Such swarms—of pigs, horses, and cattle—spread over much of the terrain in the Americas and Australia. The most dramatic swarm involved the cattle released on the Argentine pampas. As early as 1619, the governor of Buenos Aires reported that an annual cull of 80,000 did not lessen the wild herds in the vicinity of the town. An estimate dating from around 1700 put the number of wild cattle on the pampas at 48 million—a herd far larger than the native bison on the Great Plains of North America. Fields in New England are often bordered by stone walls; on the pampas the preferred material was cow skulls. From the sixteenth century until the early nineteenth, most cattle in the New World were probably wild.[9]

All the resultant grazing, rooting, and browsing imposed enormous new pressures on landscapes that had assembled themselves under very different circumstances. Pigs, for instance, are omnivores. They will eat not just whatever Winthrop meant by "groundnutts," but roots, bark, many herbs, eggs, grubs, lizards, and nearly anything else that doesn't get out of the way. Voracious and adaptable, the pig had no native analogue in the landscapes of eastern North America or Australia, or on the many islands where they were often "seeded" by mariners to insure a supply of meat during the next voyage. (See Chapter 5.) Herds of wild pigs found all these regions to be easy pickings, and in all of them, some herds continue to frustrate eradication efforts.[10]

Even where the exotic livestock had an obvious native precursor, the introduction could still precipitate enormous change. On the North American Great Plains, for example, the slaughter of the bison and the establishment of the cow as the primary grazer profoundly disturbed the native grassland. What can a cow do to a blade of grass that a bison cannot? Bison ranged over hundreds of kilometers, following the fires that swept the plains in a natural cycle of renewal, bringing on flushes of new growth. But ranchers generally want their cattle closer to home. So a patch of range that once went years without feeling the teeth of the bison tended to get chewed over by cattle every year. And the constant grazing began to rearrange the floral mosaic: the plants the cattle liked best (and those least able to recover from their attentions) began to disappear, while the plants they liked least tended to spread. The overall effect was a reduction in floral diversity. Farther west, in the intermountain region and in the Southwest, the bison had rarely intruded; the bunchgrasses native to these regions rapidly gave way to the cattle and have never recovered.[11]

Grazing pressure often favors tough exotic plants at the expense of the natives. Many of the world's most widespread rangeland plants owe at least some of their success to this mechanism. The cheat grass invasion of the western United States, for instance, was catalyzed by the heavy overgrazing of the native range. The region's ranchers, incidentally, are still reluctant to condemn cheat entirely. Cheat is wretched forage most of the time, but in early spring its strong flush of growth provides lush grazing well before other grasses are available. (That lush early growth is the product of a root system that outcompetes the native wheat grasses for water—another reason for cheat's dominance.)[12]

Since few grasslands outside Eurasia could take the uninterrupted grazing for long, the colonists began introducing European forage plants, which were better adapted to the herds. Unfortunately, any plant with the right stuff to make good forage is also likely to be a good rangeland invader, almost by definition—it thrives under heavy grazing pressure. More subtle mechanisms were often at work as well. For instance, clovers and certain other preferred forage plants can "fix" nitrogen: microorganisms growing in their roots convert elemental nitrogen, from the air in the soil, into organic compounds that plants can metabolize directly. In areas where few native plants can fix nitrogen, the process boosts the nutrient levels of the soil, and that allows in exotic weeds that would not otherwise be able to establish themselves.[13]

This impulse to "improve" rangeland by introducing exotics is still a standard part of agricultural psychology, despite the hundreds of invasions that forage introductions have already unleashed. Among the plants currently traveling the globe as forage, for example, is a member of the bean family, *Glycine wightii,* recommended on the Web site of the Oxford Forestry Institute as a tropical forage crop. One of glycine's virtues, according to the description, is that it is "easier to propagate" than kudzu, the infamous Japanese vine introduced into the U.S. Southeast for forage and erosion control. Kudzu is now one of the region's worst and most durable weeds. Glycine itself is already a pest in Australia.[14]

The herds have unleashed invasions on a microbial level as well. Rinderpest, for instance, is a cattle disease of Eurasian origin that was accidentally brought into the Horn of Africa in 1890, in the blood of infected cattle. The Masai—the region's preeminent herding people—speak of the first rinderpest epidemics in a way that recalls the 10 plagues of Egypt in the Book of Exodus: "Rinderpest was the first catastrophe, and it started like this. First of all there was an eclipse of the sun and it took place at about five o'clock in the afternoon.... It was then that the rinderpest attacked the cattle. The epidemic finished the Masai cattle."[15]

It nearly finished a lot of wild herds as well. For 70 years of so, the epidemics swept through buffalo, giraffe, wildebeest, eland, kudu, and many other species. Mortality in some species is thought to have reached 90 percent. Cattle vaccination programs eventually broke the grip of the virus, although there are still occasional outbreaks. And rinderpest has left an enduring mark on the continent. Throughout most of this century, the epidemic dynamic—the sudden population crashes followed by equally explosive recoveries— is likely to have been the main force in shaping the wild herds. And through them, it has shaped the natural communities in general. The fortunes of predators like lions and hyenas, for example, followed those of their prey. And the herbivore population crashes opened the savannas to the spread of native acacia trees from adjoining woodland. (Heavy grazing usually kills acacia seedlings.) In the East African grasslands, the hypnotically regular, even-aged stands of acacia probably owe their existence to rinderpest. In effect, the virus sculpted much of East Africa's rangeland.[16]

Among the other livestock diseases moving into African wildlife, perhaps the most serious is bovine tuberculosis, which has invaded South Africa's Kruger National Park, the "crown jewel" of the country's natural areas. In the past couple of years, the disease has spread from infected buffalo into populations of kudu, lions, cheetahs, and baboons.[17]

Agriculture is stirring up disease cycles on other continents too. Brucellosis (a cattle disease that causes spontaneous abortion) and bovine tuberculosis have infected North American elk and bison. In the United

States, hundreds of bison are killed every year out of fear that they will reinfect the western herds, which have been certified as brucellosis-free. (Wildlife biologists generally dismiss that prospect as unlikely.) In South America, the quixotic attempts to convert Amazonian forest into ranches have yielded—besides ruined forests and degraded soil—sickly cattle that may have transmitted their ailments to the continent's wildlife. Foot-and-mouth disease and bovine rabies have been cited as a possible cause of several South American mammal species declines. And recently, a poultry disease appears to have spread to Antarctica, where it has infected Emperor and Adélie penguin populations. The pathogen, known as infectious bursal disease virus, already has a worldwide distribution and a high-virulence strain of it has recently emerged. Its ecological effects in Antarctica cannot yet be predicted.[18]

Agriculture has even had its own version of germ warfare. On the plains of the western United States, ranchers thought they saw an adversary at least as formidable as the great nations of the Sioux, but marshaled somewhat lower in the grass. Prairie dogs, a group of native rodents, graze patches of prairie down hard to build their huge "towns"—warrens of tunnels topped with earthen lookout mounds. Burrows, mounds, and bare earth, often further worn by dust-wallowing bison: it sounds like some sort of natural analogue to a trailer park. But the prairie dogs are like the plains fires; what looks like damage is actually part of a healthy dynamic of renewal. Prairie dogs rarely remove more than 7 percent of the forage from an area, and the new vegetation that springs up around the towns tends to be much more diverse and nutritious than uninterrupted prairie. That is why the native

prairie grazers—the enormous herds of bison, elk, and antelope—depended so heavily on these islands of rodent business in the vast seas of grass.[19]

The ranchers, however, rarely saw farther than the bare ground. They killed as many prairie dogs as they could, and they tried to ignite epidemics by introducing sick animals into healthy colonies. But they might have stuck with rifles and poison if they had realized why the animals were sick. Bubonic plague, the infamous Black Death of medieval Europe, found a second home in the prairie dog tunnels. By the 1930s, 98 percent of the prairie dogs had been eliminated and the survivors are still being poisoned—usually at taxpayers' expense. (In some states, it is actually illegal not to kill them.) And the plague continues to flourish. It has spread through 34 species of rodents and 35 species of fleas. It has crippled surviving prairie dog populations, and since prairie dogs are a "keystone" species in the plains ecosystem, the plague is likely to be a major obstacle to ecological recovery. To humans, the New World strain of the plague is apparently far less virulent than its medieval European precursor, but it still reaches out and kills a few unlucky people from time to time. For wildlife in general, the plagues unleashed by agriculture may ultimately be as dangerous as simple brute conversion of habitat.[20]

But these invasions of weed, beast, and pest into previously natural ecosystems are really only half the story, since the ecosystems generally "invade back." As agriculture expands into new landscapes, some of the original occupants of those landscapes expand back into it: invasion provokes counterinvasion. The European corn borer is a counterinvader; so is the Colorado potato beetle, a native of the U.S. Southwest. The beetle used

to live unobtrusively on a native plant called sandbur, which is a member of the large and widely distributed potato genus, *Solanum*. (A genus is a group of closely related species.) About 140 years ago, potatoes were introduced into the region, and the beetle discovered that this relative of its native food plant was good eating as well. It has since become the most destructive insect pest of potatoes in both North America and Eurasia. This process remains a part of the agricultural dynamic, particularly in the tropics, where large-scale farming is still encroaching on large tracts of intact wetland and forest. Cutover rainforest, for example, often bequeaths destructive fungi or other pathogens to the crops that succeed it.[21]

A pest may enter the system several times or in several forms, thereby defeating hopes of reducing it to some sort of predictable equilibrium state. This is the career path of the late potato blight, originally a fungal pest of a wild potato species growing in central Mexico. The potato was brought from the New World into Spain in the sixteenth century and eventually became a mainstay of northern European agriculture. It is not clear how the fungus escaped its original range, but it spread to Europe in the 1840s and provoked the Irish potato famine, in which 1.5 million people died. The breeding of resistant varieties and the development of fungicides eventually suppressed the blight, and the potato went on to become the world's most valuable— and most pesticide-dependent—noncereal food crop.

But the blight has come out of retirement. Another strain of it emerged from central Mexico in the 1970s and had spread throughout most of the world's potato-growing regions by the late 1980s. This "A2" strain is currently turning potato fields to pulp throughout

North America and northern Europe. And it's not about to be poisoned out of the fields, since its success seems to be tied in some way to both strains' growing resistance to metalaxyl, the chemical most commonly used to suppress the blight. (Pesticide resistance is discussed later in this chapter.) More ominous still is the interaction between the strains. When they encounter each other, they can interbreed, and this sexual reproduction permits a much more rapid shuffling of genes than the nonsexual, clonal development that each strain pursues when on its own. Because of this heightened genetic activity, the blight's sexual adventures greatly increase the possibility of new strains emerging—strains even better adapted to local conditions than their parents. The blight is now reproducing sexually in both Europe and North America, and currently causing losses on the order of 20 percent of global potato production.[22]

* * * *

The invasion dynamic seems to be a permanent feature of large-scale agriculture. The system contains so many potential sources of disturbance—to both agricultural landscapes and to adjoining natural ones—that the prospect for long-term stability in either is a highly contingent affair. Consider these four dimensions of instability.

First, the boundary between crop and weed is porous. Many crops are members of "complexes"—groups of closely related species with very similar habitat requirements that often interbreed. For example, all 12 species in the oat genus, *Avena,* will interbreed—including domesticated oats and the extremely aggressive wild oats. Actually, the only reason that oats are a crop at all is that ancient Mediterranean farmers gave up treating them as weeds and started growing them intentionally. (This proved a useful strategy—many major vegetable crops, including radishes, lettuces, beets, and leeks, probably began their agricultural careers as weeds.) Interbreeding within a complex may allow a newly introduced crop variety to pass some of its genes on to the crop's wild relatives—a form of genetic invasion. Once in a while, such an invasion will result in a stable hybrid—a new weed that can then invade the fields where its domestic parent is grown. In the potato complex, for example, this process produced the Bolivian weed potato. A more common kind of leakage involves disease. Wild members of a complex can serve as alternative hosts to crop pathogens. Wild oat populations, for example, can be a reservoir for oat rust.[23]

Second, several pest species may act together, as a kind of "super-organism." In the natural world, creatures that have lived with each other, usually over a generous expanse of time, often form very close relationships. The powerful currents of biotic mixing within global agriculture are producing all sorts of new relationships at what is, from an evolutionary point of view, lightning speed. The process is essentially random: agriculture is continually sorting and resorting the denizens of its croplands, and every once in a while, two or three of them lock together. A disease finds a new vector, say, or a new "mutualism" develops—a relationship that benefits both of the organisms involved. In Hawaiian pineapple fields, for instance, the South American gray pineapple mealybug is a major pest because it helps spread a wilt disease. The mealybug is now being tended by another exotic, the big-headed ant from Africa. (In exchange for the mealy-bug's secretions, the ants chase off the bugs' predators.) Once a new mutualism like this is established, natural selection may then mold it into a more efficient form. Selection pressure may, for example, produce a higher level of protectiveness in the ants.[24]

Third, movement between the agricultural and the natural landscape can produce counterintuitive results that may damage both environments. Take the relationship between tropical rainforests and tropical crops. Diseases in banana and coffee plantations, for example, are often caused by some pathogen normally present in a nearby forest, but not at levels high enough to do the forest serious harm. Once it is within the more uniform plantation, however, the pathogen may find conditions so much to its liking that it achieves a kind of "critical mass." It becomes an epidemic, which chews up the plantation and then reinfects the forest, destroying trees that had resisted it at lower levels of infection. And these epidemics, in turn, may trigger huge new managed invasions of the crops. For more than a century, Central America's banana industry has been periodically overwhelmed by an assortment of virulent diseases. The region's banana growers now own thousands of hectares of undisturbed forest, partly as insurance against future epidemics: should a plantation be ruined, the forest can be sacrificed to start afresh.[25]

The fortunes of the honeybee in North America offer a more subtle and long-term example of this kind of reverberation. The first hives of this Old World insect arrived in the colonies in the 1620s; honey was a sought-after commodity on a continent that had little to offer besides maple syrup as a sweetener. But this invertebrate livestock strayed as readily as its vertebrate counterparts, and the bee was soon in the vanguard of colonial expansion. The indigenous people called it "the white man's fly," and much as they too learned to value the honey, they saw its arrival as a foreboding harbinger of the colonists themselves. Today the honeybee is valued not so much for its honey as for its pollinating power. Some crops rely on wind to carry their pollen—grasses like corn and wheat are wind-pollinated. But among crops that depend on insects for pollination, the U.S. Department of Agriculture estimates, perhaps somewhat generously, that the honeybee does the job 80 percent of the time. One reason for the honeybee's importance is that so many of the continent's native insect pollinators—its native flies, moths, wasps, beetles, and bees—have been fading from the scene as their habitats are destroyed and, perhaps, as the honeybee has displaced them.[26]

But the honeybee's ubiquity is now working against it. Several diseases, combined with two parasitic mite species that invaded the continent in the 1980s, have forced the bee into rapid decline. (The mites apparently arrived in the United States from Europe, the honeybee's native range, where they are inflicting similar damage.) The honeybee—in a sense, the most important animal species in U.S. agriculture—cannot be protected by quarantine because it exists in continuum over the continent's landscape. Or it did, anyway: since 1990, the country's total honeybee population (both the wild and domestic colonies) dropped by at least a quarter, and in many areas the wild populations appear to have been nearly eliminated.[27]

In the Deep South, the problem is likely to be complicated by an overlapping invasion of the Africanized "killer bee," a descendant of an African honeybee strain accidentally released from a Brazilian experimental apiary in 1957. Since then the swarms have been working their way north, hybridizing with European colonies as they go. (This interbreeding inspired the term "Africanized," to distinguish the hybrid from the pure African strain.) The Africanized bee crossed into the United States in 1990 and has thus far spread into the extreme Southwest, from Texas to southern California. It may eventually spread through most of the southern states. Although the Africanized bee will not contribute directly to the honeybee decline, it will make beekeeping—whether for honey or for pollination—more expensive. The interbreeding with European bees will force beekeepers to replace the queens in their hives more frequently, to prevent the hives from developing aggressive traits.[28]

The fourth and perhaps most important dimension of instability is the fact that pests keep changing. In agriculture as in the natural landscape, the processes of natural selection are continually at work. A crop pest population is an implacable, relentless genetic machine, performing millions of tiny evolutionary experiments simultaneously, in the bodies of its individual members. In the wild, a successful bit of adaptation may not mean very much. An insect may grow marginally better at digesting a certain type of plant; a plant may become marginally less edible. But in global agriculture, a little adaptation can rearrange the whole landscape. When a strain of the late potato blight, for example, finds a way to eat blight-resistant breeds of potato, it gains the keys to a kind of global dominion.

The most disruptive feats of adaptation usually come in response to pesticides. Precisely because they are so potent, pesticides invite evolutionary innovation. Attack a pest with a chemical that kills 99.9 percent of its population, and you offer a huge reward to the 0.1 percent that survives. Do this year after year, on field after field, and you may end up creating a monster. That is what happened, for instance, with a tiny sap-sucking insect variously known as the sweet potato or tobacco whitefly. Originally from the Mideast or Central Asia, the whitefly had spread widely throughout the globe by the 1920s and it eventually developed a taste for a huge range of crops—tobacco, cotton, melons, and tomatoes, to name a few. But it was generally a minor nuisance until it began to acquire pesticide resistance in the early 1980s. A decade later, it exploded across California, destroying tens of millions of dollars in crops in the process. And as the pesticide-resistant form—the "B-biotype"—spread around the world, its resistance strengthened, as did its appetite. Today, the B-biotype is resistant to most common insecticides and is known to attack some 600 plant species (but not, thankfully, any grains). It is also transmitting at least 60 plant viruses. The whitefly may be doing its worst in South America, where the spread of viruses by the B-biotype has led to the abandonment of more than 1 million hectares of cropland. Some experts now regard the whitefly as the world's most serious agricultural pest.[29]

The weeds are changing as well. We do not really understand what we are doing when we use herbicides to kick the genetic machinery of a weed population, but we do know, from painful experience, that this machine too is booby-trapped. Many aggressive weeds have shown an ability to develop "cross-resistance"—a reaction in which exposure to one herbicide produces resistance not just to that chemical but to other, completely different chemicals as well. (The general explanation for this phenomenon is that the weed has adjusted parts of its metabolism that many herbicides attack.) Australian wheat fields, for instance, are commonly infested by a weed called annual ryegrass. In 1977, Australian farmers went after the ryegrass with a new herbicide, diclofop-methyl, because it had developed resistance to an older chemical. By 1981, the grass had bounced back, but not just with resistance to diclofop-methyl: one carefully studied population was found to have developed some degree of resistance to 14 other herbicides. This cross-resistance is now common in Australian ryegrass infestations. And scientists have found resistance to at least one other major herbicide as well: glyphosate, sold under the trade name Roundup. Herbicide-resistant ryegrasses are popping up in North America too. If we cannot find a way to ease the chemical pressure on annual ryegrass, we may need to take a cue from the ancient Mediterranean farmers with their oats—and figure out how to eat it.[30]

"Ecological narcotics" is the term some researchers use for pesticides, and there is no doubt that agriculture has a massive chemical dependency—about $31 billion a year is spent on pesticides. But as far as we can tell, pesticides have never completely eliminated a single bug or weed. And more and more pesticides are losing their punch in one fight or another, as resistance continues to spread. (See Figure 2–2.) Currently, some degree of resistance has been discovered in about 900 pest species.[31]

★   ★   ★   ★

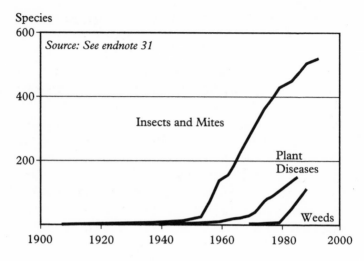

FIGURE 2–2. *Number of Pesticide-Resistant Species, 1908–94*

While particular pests may fade from the scene for one reason or another, no category of pest shows any sign of disappearing, and many are doing as well as the crops. Of the major arthropod pests (the crop-eating insects and their relatives), about two thirds are now cosmopolitan—that is, they occur virtually everywhere there is available habitat. Nearly all major grain pathogens are cosmopolitan as well. And although the evidence is much less comprehensive, some experts suspect that during the past couple of decades outbreaks of crop viruses have increased in both severity and frequency. Overall, one of the grand paradoxes of agriculture is that, despite an enormous escalation of investment in pest control, there may have been a slight long-term increase in the proportion of crops lost to pests. Pests are thought to have claimed about a third of the harvest in medieval Europe; today their take of global production probably amounts to something on the order of 35–40 percent. (See Table 2–1.) And that represents a monstrous increase in absolute terms, given the huge increases in yields since the middle of the century. (Since 1945, global yields have risen two- to fourfold, depending on the crop.) During the past 40 years, crop loss to insects alone has nearly doubled, despite a 10-fold increase in both the amount and the toxicity of insecticides applied.[32]

So perhaps it is fair to say that the mammoth yield increases have been paid for, in part, by the mammoth losses to pests. It is possible that new agricultural tech-

Table 2–1. Share and Value of Agricultural Production Lost to the Major Pest Categories, by Region, 1988–90[1]

Region	Share of Loss (percent)	Annual Value of Loss (billion 1990 dollars)
Europe	28	16.8
North & Central America	31	23.0
Oceania	36	1.9
Russia	41	22.0
South America	41	21.8
Asia	47	145.3
Africa	49	12.8
Average/Total	42	243.7

[1]Eight principal food and cash crops. Annual averages from 1988 through 1990.
SOURCE: E.-C. Oerke et al., "Conclusions and Perspectives," in E.-C. Oerke et al., *Crop Production and Crop Protection: Estimated Losses in Major Food and Cash Crops* (Amsterdam: Elsevier, 1994), as adjusted in George N. Agrios, *Plant Pathology*, 4th ed. (San Diego, CA: Academic Press, 1997).

niques may allow us to push the losses down somewhat, especially where they are well above the global
average. Africa, for example, loses about 49 percent of
its major crop production to pests, although these losses have relatively little to do with resistance, since pesticide use there is less prevalent than in other parts of
the world. (Global resistance trends, however, should
serve as a warning against trying to "fix" African agriculture by pouring on the pesticides.) And yet, taking
the system as a whole, several developments should dis-
pel any complacent faith in agricultural progress.[33]

There is, first of all, the constant dribble of new invasions. Some 70,000 pest species are already attacking
crops, but new ones are entering the system all the time
and it is never certain where they will end up. In
Central America, for example, a species of leafhopper
has been found to be carrying three extremely potent
corn pathogens—a virus and two simple unicellular
organisms called mollicutes. Outbreaks of the disease
complex in Nicaragua in the mid-1980s cut yields
60–100 percent. Will this disease complex eventually
make its way into the U.S. Midwest, the world's premier corn-growing region?[34]

And among the new invaders, there is always the
potential for something unexpected, something more
than simple additive pressure. Consider this little pink
sluglike thing—a soil flatworm native to New
Zealand—that was discovered in the British Isles in the
1960s. It has recently proved to be an extremely effective predator of the local earthworms. When it finds a
worm, the flatworm secretes enzymes that first appear
to anesthetize it and then "digest" it into a little wad of
pulp, at which point the flatworm eats it. In some
infested areas, especially in the cooler parts of the

islands, the flatworm has now essentially wiped out the earthworms. Since earthworms are critical to the life of the soil, the flatworm might eventually inflict a far more serious biotic injury on its new range than would be expected from a standard crop pest—even a fairly aggressive one. And there is always the possibility that the flatworm could spread throughout northern Europe, or that some of its relatives could find room on the continent as well. (New Zealand has about 200 flatworm species—Europe only has 6.)[35]

Several trends in developing-world agriculture may give pests that are already in the system greater room to maneuver. Since the mid-1980s, the U.S. Agency for International Development (the U.S. government's foreign aid service) and some international development agencies have been funding the production of nontraditional export crops in Latin America. The region has seen a huge surge in high-value produce like snowpeas, tomatoes, melons, mangos, strawberries—even flowers. But these export operations tend to make intensive use of pesticides, and the crops themselves are very inviting to whiteflies. That is not a good combination: heavy pesticide use eliminates predatory insects that can help keep pest populations under control, but it doesn't much bother the whiteflies themselves. As the populations of whiteflies and other virus-carrying insects explode, they are injuring not just the export crops but subsistence crops as well. Beans, for example, are a staple throughout the region, and the principal limit on bean production in Latin America is a whitefly-transmitted pathogen called the bean golden mosaic virus. Bean yields are already very low; representative yields in Central America are just under a third of what would be regarded as a solid harvest in the United States, and the number of viruses attacking beans is rising. Such developments cast a long shadow, since similar nontraditional crop strategies are being adopted in parts of Africa and Asia as well.[36]

Traditional agriculture is retreating on another front too. Throughout the developing world, farmers are abandoning their traditional crop varieties, known as "landraces," in favor of the Green Revolution strains developed by industrial agriculture. Most major crops originated in the tropics or the Middle East, and most of their genetic diversity remains in their areas of origin. The landraces are the principal vessels of this natural wealth. So, in a sense, the displacement of landraces by the Green Revolution strains is bringing the invasion ecology of agriculture full circle: the industrial varieties are reinvading their native ranges—and displacing their ancestral stock.[37]

The loss of the landraces sets the scene for further pest trouble in two ways. In the first place, it causes an immediate increase in crop vulnerability. Landraces are generally rather low-yielding in comparison to their Green Revolution cousins when the latter are grown under optimal conditions. But because landraces are so much more diverse genetically, landrace agriculture is a very durable enterprise. A traditional potato field in the Andes, for instance, may have 20–30 different kinds of potato in it. No single pest or bad stretch of weather is likely to destroy the entire crop. Green Revolution varieties, on the other hand, are high-performance but very delicate instruments: they require lots of water, chemical fertilizer, and pesticides. Even then, it is not always possible to make them work: the arrival of genetically uniform wheat in India, for example, transformed a minor fungal disease called karnal bunt into a major pest. It reached epidemic proportions in India, and has since spread into the southwestern United States.[38]

The second problem concerns the long-term security of the industrial strains themselves. Landraces are genetically dynamic; they are constantly testing themselves against their surroundings, constantly evolving new forms of stamina. An industrial variety, on the other hand, is static; its genetic composition is set by the company that produces it. To make the industrial varieties less vulnerable to pests, breeders rely continually on the genetic "wisdom" of the landraces, either directly or through seedbanks. Over the long term, therefore, the loss of the landraces could make it increasingly difficult to keep industrial varieties ahead of the pests. At present, this loss may be reducing the reservoir of crop genetic diversity by as much as 1–2 percent per year.[39]

It is true, however, that agriculture is developing new pest-fighting abilities. One of the most powerful currents in biology today involves biotechnology—techniques for manipulating genes and for moving them from one organism to another, often very different organism. These technologies hold enormous promise on many fronts—in agriculture, medicine, and even the fight against exotics (see Chapter 9). Biotech has already proved itself as a means of making far more effective use of a landrace's genetic wisdom. Several years ago, for instance, biotech researchers were able to pinpoint a gene that confers resistance to leaf blight in a strain of wild rice, and splice that gene into a domestic rice variety. (Leaf blight is the most common bacterial plant disease in the world.) That task might have taken decades with traditional breeding techniques. Biotech may eventually allow the gene to be used in many crops besides rice—an application that would not even be possible with traditional breeding.[40]

But despite its power, biotech is no more a permanent "fix" for the invasion dynamic than any agricultural technology that has gone before it. Any technology can boomerang. One focus of biotech, for example, has been engineering crop varieties with a higher tolerance to certain herbicides. (Some herbicides cannot be used effectively on certain crops, because they would kill the crops as well as the weeds.) Other transgenic crops are being developed to produce their own insecticide, generally the toxin secreted by the soil bacterium *Bacillus thuringiensis* (Bt for short). (Both herbicide-tolerant and Bt toxin crops are now in production; see Chapter 8.) In the case of herbicide-tolerant crops, it has already been shown that the engineered genes can sometimes "escape" into a crop's weedy relatives through interbreeding, thereby conferring a degree of herbicide tolerance on the weeds.[41]

This kind of genetic pollution is not likely to have much ecological importance outside the areas where herbicide is being applied. Beyond the reach of the herbicide, plants with the gene would probably not enjoy any advantage over plants that lack it. Where no "selection pressure" favors it, the gene will grow rarer with each generation and eventually just disappear into the background static of the weed's genome. But near cropland, regular herbicide exposure could spread the escaped gene through the weed population and create yet another resistance headache.

Even in the absence of such genetic pollution, widespread use of herbicide-tolerant or insecticide-producing crop systems may increase the rate at which pests become resistant. That would be especially unfortunate in the case of Bt, which is widely regarded as far less dangerous ecologically than synthetic pesticides. Resistance to Bt has already been identified in the diamond-back moth, a worldwide pest of the mustard family, which includes rape, cabbage, broccoli, and various other mainly vegetable crops. Widespread resistance to Bt would deprive farming of one of its most reliable controls against insect pests in the butterfly and moth family.[42]

It is also possible that an engineered organism itself might prove unpredictably disruptive. Such an event is very unlikely, but an experiment in 1994 showed that it cannot be dismissed entirely. Two scientists at Oregon State University were testing the behavior of an engineered microbe they hoped could be used to boost energy efficiency on the farm: they had altered a common soil bacterium to produce ethanol, a fuel alternative to gasoline, from crop residues. But the scientists found that when the bacterium was introduced into their soil chambers, the level of mycorrhizal fungi dropped by more than half. These fungi grow in close association with the roots of many kinds of plants. They play a critical role in plant nutrient intake and are essential to the functioning of ecosystems—both agricultural and natural—all over the world. "So if the bacterium had been released, it could have been a real problem," one of the scientists explained dryly to *The Oregonian* newspaper. "If the organism survived readily and spread widely, very likely we would be unable to grow crops without a control measure for this organism."[43]

Transgenic crops are a long way from the first tentative sowings of edible wild plants 10 millennia ago, and yet in a sense, both these sowings and all the harvests in between are of a piece. They are all, in one way or another, efforts in managed invasion—combined with efforts to manage the "revenge invasions" that inevitably result. Modern agriculture is a highly integrated enterprise with a global reach, and a consequent degree of biotic mixing that has little precedent in terms of either its rate or its scale. Today, it can be difficult to know what to call exotic and what native—even the native pests have become exotic in a way, because the landscapes they recolonize have been so thoroughly remolded. The dull, comfortable familiarity of a cornfield, say, or a sugarcane plantation makes it difficult to see large-scale agriculture for the radical evolutionary experiment that it is. But agriculture has dissolved ancient, stable, and highly diverse landscapes all over the world and replaced them with new, relatively simple, and extremely unstable combinations of plants and animals. In the years ahead, one of the greatest challenges that agriculture will face is increasing the stability of the artificial ecosystems it has created.[44]

3

The Forests

After all the blood and iron he had lavished on his new realm, Menelik the Second was not about to be defeated by lack of wood. By the early 1890s, after two decades of fighting the Egyptians, British, Italians, and various local clans, Menelik had managed to forge a more-or-less united Ethiopia, with himself at the head of it. But now his capital, Addis Ababa, was running out of fuelwood. Wood was as much the stuff of life as water or teff, his subjects' favorite grain. Wood meant shelter, warmth, and food. Addis had been founded just a few years earlier as Menelik's imperial seat—"the new flower," its name meant. But already it was taking on the discouraging aspect of its predecessor, Entoto, a squalid highland camp that had been abandoned for lack of wood. The landscape around Addis was barren now too. And all over the country, the acacia, fir, and scrub were disappearing. Around many towns hardly a tree was left in sight, and off beyond the blazing horizon, nomads and wood haulers chopped away at the receding bush.[1]

Bans on wood cutting meant little to his subjects, so Menelik appealed to the technology of his French allies. In its colonies on the other side of North Africa, France had extensive experience with the introduction of exotic trees. At Menelik's request, a French railway engineer arranged for the introduction of a species of eucalyptus, a group of trees native to Australia and some of the nearby islands. *Eucalyptus globulus,* the blue gum tree, arrived in Ethiopia.[2]

Menelik proceeded to conjure up a huge gum forest. He gave the trees away or sold them for next to nothing; he exempted land planted in eucalypts from taxes. The sharply perfumed "incense trees" even took on the odor of sanctity, when the Ethiopian Orthodox Church, ascendant over matters spiritual in Menelik's realm, endorsed the planting too. By the first decade of this century, Ethiopia's chronic wars had become a distant clamor in a mind confused by strokes and syphilis, but Menelik could watch his blue gums gathering strength. Tufts around the houses began to merge into a blue-green sea that floated above the dust.[3]

Less than a decade after Menelik died in 1913, Charles Rey, an Englishman who spent several years in Ethiopia, described the road to Addis from the north. From the furnace of the Blue Nile Gorge, the traveler ascended a steep, crowded path and emerged into the cool air of the Shewan highlands, where "one seems to be about to enter a forest, and it is only on a near approach to the town that houses begin to stand out amongst the trees, and the rays of the sun sparkling and glittering on the metal roofs and white-washed walls make one realize that a town, and an extensive one at that, is hidden in the foliage."[4]

Addis Ababa, the administrative and industrial heart of Ethiopia, was built on the vigor of the blue gum—as were hundreds of Ethiopian villages. Its urban gum forest was so extensive, and so much a part of Addis, that a French contemporary of Rey's suggested changing the city's name to Eucalyptopolis. In most of Ethiopia today, the native forest cover is a minor part of the landscape.[5]

There is something primal in Menelik's achievement: getting the landscape to do more or less what you want it to is perhaps as close to magic as humanity is likely to get. We are used to seeing this happen with crops, but we recognize agriculture as an industry—with all the material and financial "inputs" that come under that rubric. With trees, the gradation between the wild and the artificial is less obvious. And Menelik's blue gum state may well be unique: probably no other economy in the world chops so much of its wood in artificial forests of exotic trees.

But that may change. Plantations of tough, vigorous trees like the blue gum are carpeting more and more of the planet, even as natural forests continue to shrink. The data are very vague, but it seems that tree plantations cover at least 180 million hectares—just 10 million hectares less than the total land area of Mexico. (And that's excluding the enormous agricultural tree plantations of the tropics, primarily palm and rubber.) About half of global plantation area is in industrial countries and half in the developing world.[6]

A great many plantations are still nonindustrial—

they consist of community woodlots, agroforestry projects, erosion control plantings, and so on. But at least 100 million hectares of the global plantation estate is for the production of timber, wood chips, pulp, and industrial fuelwood. And this area is expanding vigorously, especially in the tropics. Between 1965 and 1990, the tropical plantation area increased at least fivefold and possibly sixfold; most of the countries with major plantations have announced plans to double their plantation areas between 1995 and 2010.[7]

North American and European plantations are composed mostly of trees native to the regions in which the plantations are installed. That does not necessarily make them environmentally benign, of course, but it may allow for a substantial degree of continuity between the managed and the wild landscape. In some areas, particularly in Scandinavia, the distinction between plantation and native forest is not always obvious. Elsewhere, however, the industrial plantation is primarily an exotic industry relying extensively on a dozen or so species of eucalypt, pine, and, to a lesser degree, acacia. In 1990, eucalypts and pines accounted for more than 80 percent of the plantation area in tropical America, about 50 percent in tropical Africa, and about 20 percent in Asia-Oceania. Virtually all these plantings are exotic, and so are many of the rest. The eucalypts and pines are marching over the warm temperate regions as well: Argentina, Australia, Chile, China, New Zealand, Portugal, South Africa, Spain, and Uruguay all have extensive exotic plantations in these species.[8]

Exotic eucalypt plantations alone covered some 10 million hectares in 1990, and their realm is likely to be considerably larger by now. Some eucalypt species, such as the blue gum, are probably now growing in virtually every country within the tropical and warm temperate regions.[9]

The eucalypts and their fellow exotics are not generally destined for wood production in the ordinary sense of the term. Probably less than 20 percent of tropical industrial plantation area is grown for lumber. Fuelwood is a more common product, both for private homes and for industrial use. In southeastern Brazil, for instance, the expansion of eucalypt plantations to fire the area's many steel mills has been a major factor in the continued decline of the country's Atlantic forests—perhaps the most biologically diverse and endangered forest communities in the world. In the early 1980s, close to 40 percent of tropical plantation area was devoted to fuelwood. But that proportion has probably declined since then, in favor of the "fiber" sector, whose main customer is the $337-billion-a-year pulp and paper industry. It is impossible to pin down an exact proportion (partly because the same trees can be used for either fuelwood or fiber), but most tropical industrial plantation area today is probably producing paper, fiberboard, or other low-grade wood products.[10]

Increasingly, the pulp and paper industry is using developing-country plantations to feed the paper addiction of industrial countries, which currently consume more than three quarters of the industry's global output. Paper and paperboard now account for about 44 percent of international trade in forest products, by value. And given the growing importance of developing-country pulp production, it is perhaps not surprising that plantations are absorbing a substantial amount of foreign aid. Pulp plantations are often funded by governments of countries such as Canada, Finland, Japan, and Sweden, which have highly developed forestry or paper sectors that stand to benefit from the plantations. (The plantations serve as sources of raw material, or as customers for the domestic industry's engineering, construction, and consulting sectors.) International lenders are also involved. Between 1984 and 1994, for example, the World Bank lent more than $1.4 billion for the establishment of 2.9 million hectares of tree plantations all over the world.[11]

Fiber plantations are a paradoxical operation: despite the furious planting of billions of trees, they are deeply implicated in the machinery of deforestation, especially in the tropics. To understand why this is so, you must first see a pulp plantation for what it is—a kind of cyborg landscape, superficially natural but really just part of a machine. Most plantations are monocultures, like giant cornfields. And, like cornfields, they require intensive management. To suppress pests and competing vegetation, they are doused with pesticides, which poison soils and watersheds. In southeastern Brazil, for instance, pesticide runoff from the CENIBRA and Bahia Sul Celulose plantations has apparently ruined the fishing base of local communities.[12]

Plantation soil is often compacted by heavy equipment and eroded by downpours that pass undeflected through the sparse plantation canopy. And then comes the harvest: pulp trees can be clearcut in 10 years or less. The cumulative result is a severe insult to the soil. Depleted of nutrients, compacted, and scarred by gulleys as deep as a meter, plantation soil is usually good for little else besides another crop of chemically dependent exotic trees—and eventually not even good enough for that.[13]

Apart from sheer displacement of native forests, the plantations exert a number of malign effects that extend well beyond the plantation boundaries. Both eucalypts and pines are widely blamed for lowering water tables. (Eucalypts especially are famous for being able to absorb large quantities of water through their roots, and then "exhaling" it through their leaves—a quality that gained them a role in many wetlands-draining projects during the nineteenth century.) In countries as diverse as Brazil, Chile, India, Spain, and Thailand, people have complained about the drying effect of the plantations; in some areas in Thailand, water tables have reportedly dropped to levels that make rice farming impossible. Eucalypt plantations owned by Aracruz Celulose, in eastern Brazil, are reported to have dried up 156 streams and one river, the São Domingos.[14]

Many commercial species are spreading beyond the bounds of their plantations and invading valuable natural areas. The Usambara Mountain range of northeastern Tanzania, for example, is a "hot spot" of plant diversity: the area contains many endemic species (that is, species that grow nowhere else). But the plantation tree *Maesopsis eminii*, native to central Africa, threatens to displace much of the native forest. At least 19 pine species have invaded various countries in the southern hemisphere—including Australia, Madagascar, Malawi, New Zealand, South Africa, and Uruguay. Eucalypts seem generally to be less invasive than pines, although they appear on weed lists in many countries, and eucalypt invasions are displacing what remains of the native vegetation in areas of the Mediterranean Basin.[15]

Some plantation species are spreading by fire. They burn readily, and the fires extend into surrounding vegetation; then they colonize the burned-over areas before native plants can come back. This mechanism is at work in South Africa's Cape Floral Kingdom, another endemic hot spot that contains perhaps the most diverse flora on Earth. (See Chapter 5.) Exotic pines, acacias, and eucalypts have burned their way into the Kingdom and now dominate thousands of hectares, threatening some 750 endangered plant species.[16]

The governments, corporations, and aid agencies involved in the plantation boom commonly claim that plantations reduce the pressure on native forests and that they constitute a kind of reforestation. By and large, these claims are a rhetorical sleight of hand. The tropical pulp operation is inimical to native forests, partly because it wants to "mine" the soil and water underneath them, and partly because it cannot resist the quick profit from logging. That is why pulp plantations are generally part of a package that includes, either directly or indirectly, the logging of native forest. In what remains of Thailand's native forests, for instance, local companies allegedly pay people to log illegally, so that the land can be recategorized as degraded, which exempts it from the country's strict anti-logging statutes and makes it available for plantations.[17]

In places where the forest is more plentiful, such pretexts are usually unnecessary. Native forest may be logged to feed new pulp mills, which will eventually operate with plantation fiber once that becomes available. The Perawang pulp mill on Indonesia's island of Sumatra, for instance, is chewing up 200 square kilometers of highly diverse old growth forest every year while waiting for its acacia plantations to mature in 2000. In such cases, logging profits cover the cost of setting up the plantations, while the plantations provide the public relations cover for the logging. A vice presi-

dent for the company that owns the Perawang operations once explained the process with refreshing can-
dor: "Basically, we are looking for forest that can be
clear-cut and replaced with eucalyptus and acacia."
Thus far, Indonesia has converted some 2.2 million
hectares of forest to plantation, and the country plans
to convert another 6 million hectares within a decade.[18]

As the cyborg landscape swallows up natural space,
it is also invading cultural space. Another big difference
between a forest and a plantation is that the latter has
precious little room for people. In many parts of the
developing world—in countries as diverse as Chile,
Brazil, Indonesia, and Thailand—indigenous and other
rural peoples, who are often not integrated into the
mainstream economy, have been forced out of the
forests to make way for the eucalypts. In such circumstances, the plantation may become a kind of autocrat-
ic instrument—a means of enforcing the will of the
state in regions that may not be inclined to obey.

There is a logic of analogy here: both forests and cit-
izenry can be "improved"—made more orderly, obedi-
ent, and productive in ways that the authorities are likely to value. In Thailand, for instance, the military junta
that seized power in February 1991 designated some
14,700 square kilometers for private-sector plantations
in the country's northeast, an area identified as a center of resistance to military rule. The program was sus-
pended when the junta was overthrown in May 1992,
but not before some 40,000 families had been displaced. In Indonesia, plantations are destroying the traditional land tenure system of the forest peoples, along
with their renewable forest economies based on fruit,
rattan, and slash-and-burn agriculture.[19]

In a few countries, popular opposition has stopped

or at least slowed the plantation invasion. "The fascist tree" is what small-scale farmers in Spain and Portugal often call the eucalypt, and in Spain especially, the label is apt: large-scale plantations are a legacy of the Franco era. In both countries, a spate of anti-eucalypt riots broke out around 1990, and several new plantations were demolished. The eucalypt seems to have gone into decline on the Iberian Peninsula.[20]

Thai farmers have taken similar action, but with more limited results. In the Thai countryside, eucalypts are sometimes called "demons" because they hog nutrients and water and they harden soil. Resistance has sputtered along for more than a decade; the biggest flare-up was a riot in Buriram province, northeast of Bangkok, when some 2,000 people burned a plantation. Opposition has forced the cancellation of some projects, as in 1990, when Shell Oil abandoned its plans to plant 12,000 hectares of eucalypt inside a forest reserve in Chanthaburi province—a project that had attracted favorable attention from both the Canadian and Finnish governments and that involved illegal logging. In Indonesia, however, numerous small rebellions against the plantations seem thus far to have achieved little.[21]

Elsewhere the invasions continue largely unopposed. In Laos, tens of thousands of hectares of plantations are being planned by Thai companies, the Laotian military, the World Bank, the Asian Development Bank, and the Finnish government. Other projects are reportedly in the works for Myanmar (formerly Burma), which has about 70 percent of the world's remaining natural teak forests. And in southern China, Indonesian plantation magnates are setting up joint ventures to put more than 220,000 hectares into eucalypt. The process has perhaps reached its extreme in Chile: 80 percent of that country's forest industry is in Monterey pine (native to southern California) and eucalypt; forest products have become the second largest export sector after copper; and according to the U.N. Food and Agriculture Organization, the rapacious destruction of its native forests now qualifies Chile as the second most deforested country in the world.[22]

*   *   *   *

Industrial forestry is hardly the only industry that is spreading exotic trees—and in terms of the number of species introduced, it is not even the most important one. Thousands of tree and shrub species are being moved from place to place in the developing world for motives more like Menelik's—from a desire to satisfy a local need. The proponents of these introductions are struggling with some of the least glamorous and most dysfunctional aspects of humanity's relationship with the Earth: lack of fuelwood and desertification, erosion, and the gradual failure of farming. Given the often dismal local conditions, it is hardly surprising that the search for solutions should so often lead to some "wonder tree" in another part of the world.[23]

A representative sample of discoveries:

- A workshop announcement on the mesquite genus: "*Prosopis:* Semi-arid Fuelwood and Forage Tree—Building Consensus for the Disenfranchised." (*Prosopis* species have invaded Australia, Hawaii, Namibia, Oman, and South Africa.)[24]
- *Mimosa scabrella*, a tree native to Brazil, "shows great promise as a multipurpose agroforestry tree in the rugged highlands of Rwanda." (*Mimosa* species have invaded the U.S. Southeast, various

Pacific islands, Southeast Asia, Australia, and Zambia.)[25]

• The shrub *Eleagnus umbellata*, native to central Asia "has great potential for agroforestry systems in temperate zones." (*E. umbellata* is a pest plant in the central and eastern United States.)[26]

The fact that these plants are known invaders does not necessarily disqualify them for any role in a managed landscape. No doubt, some commonly planted exotics will continue to be the best answer for certain types of waste-lot reclamation or home fuelwood production. But a record of invasive behavior is a clear signal to treat the plant with caution.

Yet recommendations such as these are almost never tempered by caution. The newsletter that printed the *Eleagnus* note, for instance, also contained a recommendation of *Parrotia jacquemontiana*, a shrub native to the southern Himalayas. The authors point out that in its native range, it is already regarded as a nuisance by foresters because its dense stands prevent regeneration of the Deodar cedar, one of the region's dominant forest trees—yet they conclude that "with work on propagation, we believe it will be an important species in agroforestry research." Even proven invaders still get promoted: an African species of acacia (*Acacia nilotica*) is being promoted in parts of Africa where it is exotic as well as in India, while officials in Indonesia and Australia are trying to eradicate it.[27]

This kind of well-intentioned, "populist" forestry has introduced at least 2,000 exotic woody plant species into various parts of the world. Thus far, at least 135 of them have launched invasions. Yet a successfully established exotic plant may sit for a few years, for a decade, or for longer than a century before launching an invasion. Those 135 species could well be harbingers of enormous ecological change.[28]

*   *   *   *

Along with the intentionally introduced trees and shrubs, forests are increasingly congested with accidents—plants that have escaped from gardens or hitchhiked in on visitors' gear, or that have followed the loggers into the forest. One of the world's worst weeds of disturbed tropical forest, for instance, is alang-alang grass, a native of South Asia. In some places, the grass is a valued source of thatch, but its career as a weed far overshadows its economic importance. Once alang-alang gets a grip on a piece of ground, it does not let go; it does not let anything else grow there, and it does not let depleted soil regain its fertility. From East Africa through Oceania, alang-alang has become a part of the somber rhythm of deforestation: a patch of forest is cut over, the grass gets hold, and the trees cannot come back.[29]

Abysmal land management is estimated to have ceded a full 20 percent of the Philippines' land surface to alang-alang; 10 percent of the original forest area of Indonesia's outer islands was alang-alang in 1987— about 15 million hectares. In Asia, alang-alang is thought to have claimed some 50 million hectares in all. The same thing is happening in the New World. During the past century, several African and Asian forage grasses have been slowly strangling various types of forest, from the U.S.-Mexico border to northern South America—moving in when they can, then waiting for the next fire or bout of logging.[30]

Alang-alang grass has numerous animal counterparts—mainly insects—in forests all over the world.

The most notorious example is the gypsy moth, a wide-spread pest of Eurasian forests, which was introduced into North America in the 1860s by Leopold Trouvelot, a French artist and entomologist. Trouvelot had brought the moth to his home in Medford, Massachusetts, in the vain hope of starting a silk industry. And through an open window, the moth found an open continent. According to the first official account of the disaster, Trouvelot knew the moth was dangerous, and "finding his efforts for its eradication futile, gave public notice of the fact that the moth had escaped from his custody." The first major outbreak came in 1889, when the moth stripped Medford and its environs bare of foliage, creating a summer version of December. It now occupies a region from southern Maine and Quebec, as far west as Michigan and as far south as Virginia. Occasional outbreaks beyond this zone may be harbingers of further invasion.[31]

As with alang-alang grass, the gypsy moth's conquest is partly a side effect of other forms of forest degradation. Logging and diseases have simplified the forests of eastern North America, eliminating trees the moth would rather not eat—while its favorite oaks have more room than before. The moth is usually at its worst on the expanding fronts of its range. In areas it has occupied for many decades, its effects may be hardly perceptible most of the time—a quality typical of other chronic forest maladies, such as acid rain. But unpredictable flare-ups occur from time to time. In 1981, the moth defoliated 5.2 million hectares, mostly in the U.S. Northeast; in 1991, it stripped 1.7 million hectares. The moth slows the growth of the trees it defoliates but does not necessarily kill them; during a severe infestation, however, tree death can be as high as 90 percent.[32]

Of course, the gypsy moth is just one of many traveling forest pests, and some of the more recent arrivals, in North America and elsewhere, may prove to be just as dangerous. There is, for instance, an insect that Chinese entomologists are calling "the white moth"— apparently an immigrant from North America that has defoliated thousands of hectares of forest in central China. Will it become North America's "revenge" for the gypsy moth? There is East Asia's longhorned beetle, which was discovered in New York City in August 1996. Will it do to North American maples what the gypsy moth is doing to oaks? (In its native range, the beetle attacks a wide variety of deciduous trees.) Nor is it just trees that are at risk from exotic forest pests. In the southern beech forests of New Zealand, there is a group of native scale insects that secrete a kind of high-calorie "honeydew" that is a primary food for other insects and many native birds. But the European wasp, which invaded in the late 1970s, now monopolizes the honeydew, throwing the long-term survival of its native competitors into doubt.[33]

One of the least-publicized aspects of global forest decline during the past century has been an epidemic of forest epidemics. These outbreaks get little public attention because they can take decades to run their course—the process often escapes individual human memory. Take the chestnut blight in North America, for example. This fungus was brought to New York City in 1904 in a shipment of Asian chestnut trees, and over the next 50 years it eliminated the American chestnut tree from its entire native range—the eastern third of the United States. The tree apparently dominated much of the original eastern forest; heavy logging in the seventeenth and eighteenth centuries probably increased its dominance because it could regenerate faster than many of its competitors.[34]

According to an old forestry "chestnut," a squirrel could travel from Maine to Georgia without ever leaving the boughs of chestnut trees. This is a ridiculous exaggeration, of course, but it is thought that at the turn of this century the chestnut may have accounted for some 70 percent of the eastern forest by basal area (the area occupied by the trunks roughly at chest height). And it is likely that the chestnut was a "keystone species," because its enormous quantities of nuts would have fed so many birds and mammals. The demise of the chestnut, combined with heavy hunting pressure, must have made the eastern forest a much emptier realm.[35]

Australia may be in the grip of an even more disruptive epidemic. A form of cinnamon fungus, a relative of the potato blight, had arrived in the country by the 1920s and spread throughout much it during the ensuing half-century. Over thousands of hectares, it killed up to three quarters of the plant species it encountered, from canopy trees to understory shrubbery. The disease, known as "the graveyard syndrome," has eliminated highly diverse forest communities, some containing hundreds of rare endemic plant species, along with their resident birds and marsupials. Weedy grasslands eventually spring up within the dead and dying stands. The origin of the fungus is unknown, but it has been discovered living in a relatively peaceful and presumably native state in several areas around the world, including South Africa and New Guinea. It is at its worst in Australia's forests, but is at work elsewhere as well. By the 1820s, it was probably present in eastern North America, preying on members of the chestnut genus; and since the 1940s, it has been implicated in oak declines in several western European forests.[36]

As with the other types of infestations, some epidemics have clearly been catalyzed by disturbance. In eastern North America, for instance, the American beech is losing ground to a fungus transmitted by a tiny insect called the beech scale, which arrived in Nova Scotia from its native Europe around 1890. Earlier generations of loggers did not think much of the beech, so when they cut these forests, they left a good deal of it in place. The open stands made it much easier for the disease to find its victims. As the beech began to die off, a round of salvage logging cleared out the larger trees in many areas, leaving only dense thickets of beech "suckers" sprouting from the roots of the cut trees. These thickets tend to suppress more diverse cover and incubate still more of the scale.[37]

But in many cases, the epidemics leave a great deal of room for guessing—about their origins, their mechanisms of spread, and their eventual impact. A pine wilt nematode appeared in Japan in the early 1900s, but has only recently exploded across Japanese forests, killing 10 million trees in 1981 alone. (A nematode is a kind of microscopic worm.) The nematode may be from North America, where it is a widespread but as yet relatively minor pest. Why did it "incubate" for so long in Japan? Will it eventually explode across North America?[38]

In eastern North America, a pathogenic fungus is killing off the butternut, a member of the walnut genus. The fungus is probably exotic, but where did it come from? Another fungus has invaded the region, perhaps from East Asia, and is suppressing a little understory tree, the eastern dogwood. The dogwood is very efficient at pulling calcium out of the soil and depositing it, through its leaf litter, on the forest floor—a process that greatly reduces the rate at which calcium is leached out of the system. Calcium is an essential nutrient for both plants and animals, so what will the decline of the dogwood mean for these forest communities as a whole?[39]

As in agriculture, the spread of a forest pest may allow it more "breeding room"—stimulating the development of new strains and creating opportunities for reinvasion. Take the fungal pathogen that causes Dutch elm disease. The disease arrived in North America from Europe in 1930 in a shipment of elm veneer logs. It has since spread throughout the eastern half of the continent and forced the American elm—the huge, vase-shaped tree that graced the streets of so many of the region's small towns—into functional extinction.[40]

The disease gets its common name from the place where it was first described, in 1921, but it is no more Dutch than American. Before it arrived in the New World, it had already decimated European elms. It probably originated somewhere in Asia; Asian elms tend to be resistant to it, which suggests that they may have evolved with it. In the decades since its North American invasion, new races of the fungus have emerged and Europe has been reinvaded from both Asia and North America. The new strains have been assigned to a new species, which appears to be displacing the original pathogen. Where exactly that species originated is not known, but its effects are depressingly apparent: the European elms have been hit hard again. Like their American counterparts, they are now little more than clumps of diseased shrubbery.[41]

*   *   *   *

Epidemics and other pest infestations tend to be seen as problems for natural forests—albeit often highly disturbed ones. Foresters, of course, are well aware that plantations too have pest problems, but they tend to view these as essentially technical glitches—mistakes that can be corrected by a better choice of tree or a change in management regime. "There seems little doubt that with proper care in selecting species, provenances, and families, risks are low," a representative opinion reads. "No major plantations have been devastated by pests or diseases."[42]

That assessment appears to ignore a great deal of evidence to the contrary:

- In China, a "witches' broom," an organism that stunts and deforms trees, has spread throughout the country's plantations of paulownia, a major timber species. Originally a minor part of the tree's disease burden, the witches' broom underwent an explosive expansion as paulownia plantations were scaled up substantially, beginning in the 1970s.[43]
- In India, eucalyptus plantations installed on more than 40,000 hectares of clearcut rainforest in the Western Ghat range were lost to a fungus that apparently emerged from the cutover forest.[44]
- In South America, a fungal disease of the guava has recently begun to defoliate eucalyptus plantations. A prominent Australian plant pathologist has voiced concern that the fungus could reach Australia and attack native stands of eucalyptus.[45]
- In Uruguay during the 1960s, virtually all of the country's recently installed Monterey pine planta-

tions were destroyed by the European shoot moth and an associated fungus. The country had to abandon Monterey pine as a forestry species.[46]

- In New Zealand, efforts to control the *Dothistroma* blight, a fungus that attacks pines throughout the tropics and southern hemisphere, involve the spraying of fungicide on some 90,000 hectares of Monterey pine annually.[47]
- In the western Pacific and increasingly throughout the tropics, plantations of *Leucaena* species, a group of fast-growing trees from central America, are being invaded and killed by various insects from the New World tropics.[48]

All major plantation trees now have serious pest problems. Eucalypts are commonly plagued by leaf-cutting ants, and increasingly their native weevils, snout beetles, and borers are catching up to them. Pines are attacked by woodwasps and aphids. Acacias are eaten by leaf cutter ants and bag worms, and virtually every plantation species is food for several kinds of fungus.[49]

The general ecological outlook is not encouraging. In tropical forests especially, almost the only pests that have been studied in any detail are ones already active in plantations or heavily managed second growth. There are undoubtedly thousands of others about which we know nothing. And since it often takes quite some time for a pest to attack an exotic tree—30 to 50 years is not uncommon among insects—initial success is no guarantee of long-term viability. Add in the tree's own native pests, which, given enough time, generally manage to track it down, and the long-term trend is fairly predictable. Brazil's eucalypt plantations, for example, were largely free of infestation at the turn of the century, but now have many pests. The same is true of Indonesia's plantations of mahogany (a tree native to South America), and in South Africa, new infestations are hitting second- and third-rotation pine plantations. The South African situation suggests that some plantations may develop a kind of double liability: an exotic tree may acquire a pest burden serious enough to injure its economic potential—but not serious enough to injure its ecological potential as an invader in its own right.[50]

In response to the growing disease burden, breeders are producing disease-resistant cultivars of some tree species. Traditional tree breeding is a much slower process than crop breeding, because trees take much longer to mature. Biotechnology is speeding the process up, but given the numbers of potential pests and the expense of replanting with new varieties, breeders are not likely to reverse the general trend. Even where they succeed, a cultivar's genetic uniformity may offer little resistance to the next pest. Plantation forestry may find itself in a kind of "red queen" predicament: like Alice in *Through the Looking Glass*, it may have to run harder and harder just to avoid losing ground.[51]

Most work on tropical forest pests has focused on infestations moving from natural forest into plantations; we know very little about the disruptive potential of movement in the opposite direction, even though it may actually be far greater. We do know, however, that this kind of movement is occurring. In India, a common pine plantation pest, the *Cercospora* needle blight, has moved out of Monterey pine plantations and is now threatening two native Indian pines, *Pinus roxburghii* and *P. wallichiana*. In Kenya and Malawi, an aphid that started its career in plantations of Mexican cypress is now attacking two indigenous trees—*Widdringtonia nodifolia*

(Malawi's national tree) and a juniper (*Juniperus procera*). In Australia, the cinnamon fungus used pine plantations as one of its routes into the eucalypt forests it killed. Tree plantations could magnify the effects of native pests as well, by generating the same kind of "critical mass" that occasionally builds up in Central American banana plantations. (See Chapter 2.)[52]

The most dangerous effects of plantation epidemics may be caused as much by finance as by biology. Today, a single pulp plant can cost as much as $1 billion, and investments on that scale put paper company executives under heavy pressure to feed the mills as cheaply and quickly as possible. That is not an environment conducive to rethinking basic premises. If a eucalypt stand grows too infected to use, any natural forest remaining in the area will look very tempting, both as a short-term wood supply and as space for new plantation.[53]

★   ★   ★   ★

A century after Menelik introduced the blue gum to Ethiopia, the union between tree and state is showing signs of strain. Addis itself is booming, but Ethiopia is slipping back into Menelik's predicament. Demand for fuelwood and lumber is probably outpacing domestic supply, although it is difficult to say for sure because the country's plantations and remnant forests are not well mapped. Even so, Menelik's strategy will probably not be repeated. In Ethiopia, too, people are suspicious of the eucalypt's greed for water. Around Addis, older people remember a time when the streams flowed year-round, but those days are gone. The country was planning to add another 3.5 million hectares to its eucalyptus plantations by 2000; the plans have apparently been set aside.[54]

Yet Ethiopia may have the makings of a vibrant forest economy. Most of the country's native forest cover has disappeared, but the native species themselves—hundreds of them—are still growing in parks and surviving stands here and there. Apparently at least 30 of them have strong potential for commercial use. So perhaps the time has come to replant what has been cut away. It is true that native forest projects may never meet all of Ethiopia's wood demand. But in the process of satisfying a major share of it, they could also provide a powerful mechanism for conserving the country's dwindling biological wealth. Ethiopia has much to gain from seeing a lot more of its native forest—and a lot less of the Emperor's new tree.[55]

4

The Waters

"The white people put the baby monsters in the lake to help us." The old woman laughed. She had spent her whole life on the shores of Lake Victoria, and must have remembered the colonial administration, so perhaps she was being ironic. Or perhaps she just thought the question was so obvious as to be ridiculous. It had come from Tijs Goldschmidt, a Dutch biologist (and fluent speaker of Swahili), who in the mid-1980s was trying to nail down the exact moment and mechanism of the monster's arrival. On the broad point, at least, his informant was right. The monster had indeed been a part of the white man's burden: the official British stocking program dated from 1962. But a bit of free-lancing by a local fisheries officer probably brought

Lake Victoria its first monster in 1954. "Every time I thought about it," Goldschmidt later wrote, "I was amazed that for the total disruption of the largest tropical lake in the world, nothing more had been needed than a man with a bucket."[1]

Lake Victoria is the second largest lake in the world (after Lake Superior) and its fisheries are a major protein supply to the three nations that cradle it: Uganda, Kenya, and Tanzania. It's a fairly young lake—just 12,000 years ago, it was dry savanna—but it has been the scene of one of the most explosive radiations of life in recent evolutionary time. A group of cichlids, the type of fish that contributes so many colorful aquarium specimens, had filled those empty new waters with a huge array of forms. The lake became a jewel box of brilliant little fish—more than 500 species, most from a single genus *(Haplochromis)*.[2]

But in commercial fisheries parlance, these haplochromine cichlids were mostly "trash fish"—too small to be worth catching. The fisheries depended largely on a couple of native tilapia species, and by the early 1950s, overfishing had pushed the tilapia into serious decline. (Tilapia are a group of large, mostly herbivorous cichlids.) The colonial authorities thought they saw a good deal of unused potential: the tilapia declines were leaving a lot of aquatic plants unconsumed. All those little fish were a waste too: they could be feeding big fish, which could be feeding people. So three exotic tilapia species were introduced from Lake Albert, to the north. Lake Albert had a remedy for the trash fish problem as well: the monster, a freshwater leviathan called the Nile perch.[3]

The perch can grow 2 meters long and weigh up to 200 kilograms. Even so, it spent the first few decades of its life in the lake quietly—a little disappointing for such a voracious predator. But during the 1980s the lake began to rearrange itself. In Ugandan waters (which amount to 43 percent of the lake), the catch shot up from 17,000 tons in 1981 to a peak of 132,400 tons in 1989. The perch accounted for more than three quarters of the 1989 catch, and the exotic tilapias for virtually all the rest. It was much the same story elsewhere in the lake.[4]

Meanwhile, Goldschmidt and his colleagues were finding fewer and fewer haplochromine cichlids, and not much else either. The native species were vanishing. Only the little native sardine was flourishing in the presence of the monster—perhaps because its high reproductive rate conferred a kind of immunity. Eventually ecologists concluded that some 200 species had disappeared completely: the greatest single paroxysm of extinction ever recorded.[5]

So many predators and so little prey: the lake's ecology appeared to have been pulled inside out. Goldschmidt described watching the process as like looking at the Serengeti plains and seeing herds of lion instead of antelope. The perch probably could not live by sardines alone, and many ecologists were guessing that it would starve once all the cichlid populations crashed.[6]

The general condition of the lake appeared to compound the perch's problems. To smoke the mountains of perch being hauled out of the water, the fishery needed huge quantities of wood. As the local rate of deforestation accelerated, more and more soil was blown into the lake. The plant nutrients in the soil combined with growing quantities of sewage, sugar mill waste, and farm runoff to conjure up mats of algae. The algae spread over the water in the absence of the algae-eating cichlids.[7]

In the murk below, many of the relict cichlid populations crashed because the fish could not see well enough to find mates. And periodically, rotting algae blooms would use up most of the oxygen dissolved in the lake. During these episodes of anoxia (severe oxygen deficiency), thousands of suffocated sardines and perch would stew in the algal mats, then sink to the bottom, where the accretion of decay robbed the water of even more oxygen. By the early 1990s, 50–70 percent of the lake's water mass was anoxic year-round.[8]

Yet it was this process that gave the perch a means of survival. Below the surface waters, in the anoxic region, certain organisms that needed very little oxygen found refuge. There, they could live off the rain of detritus, secure from the predators in the oxygenated waters overhead. These creatures—prawns, snails, and midges—underwent a population explosion. Then huge quantities of adult prawns moved into the oxygenated zone and became a main item on the monster's menu. The perch had recovered from the loss of the cichlids.[9]

From the west, down the Kagera River and into the lake's warm, fetid waters, floated lettuce-like rafts of green festooned in gauzy purple bloom. Around 1990, the water hyacinth arrived and one monster met another. The hyacinth, a native of the Amazon Basin, is thought to have been brought to Africa as a pool ornament in the nineteenth century. It was soon decorating much of the continent. Water hyacinth is one of the world's fastest-growing plants and probably the world's worst tropical aquatic weed. (See Table 6–2 in Chapter 6.) In the absence of the insects and fungi that evolved to feed on it, and in the presence of enough nutrients, it *becomes* the landscape.[10]

Like the algae, the hyacinth found Lake Victoria's waters ideal. Hyacinth likes shallow water, and by 1996 it was strangling 90 percent of the lake's shoreline. The shallows are where fish breed—the exotic tilapia and perch as well as any remaining haplochromine cichlids. But the hyacinth is choking out the light and sucking out the oxygen; even the shoreline has become anoxic. And the hyacinth's high respiration rate—the rate at which it "breathes" water into the air—is dropping the lake's water level. It has choked off the harbors too; fishing boats cannot punch through the thick mats of snake-infested hyacinth to reach open water. Even large craft are sometimes marooned in hyacinth for days.[11]

By the fall of 1997, tens of thousands of fishing families had lost their livelihoods to the weed; many had abandoned the lake and moved into the cities. The hyacinth is also clogging the pumps that supply water to the city of Kampala. It is threatening the region's major generating station at Owen Falls Dam, on the coast of Uganda. And it is incubating snails and mosquitoes. Bilharzia, a disease caused by a blood parasite, is transmitted by the snails, and malaria by the mosquitoes. In the lake basin, both diseases are already on the upsurge.[12]

The Lake Victoria basin is home to some 30 million people and has one of the fastest growing population rates in the world—about 3 percent a year. The region cannot live without its fishery—in Uganda, for example, fish accounts for half the national protein supply. But the hyacinth has conquered spawning grounds and fishing piers; the perch itself has chewed the lake's food web down to a few overstressed strands. Most experts foresee an imminent decline—if not a collapse—in the catch. Even the U.S. Department of Defense, not exactly a hotbed of environmental alarmism, is concerned about the consequences. A 1995 report by the U.S. Defense Intelligence Agency (DIA) predicted that a crippled fishing industry in the lake "would pose a greater humanitarian threat to [East Africa] than any of the recent famines caused by lack of rainfall or desertification." By way of analogy to Ethiopia and Sudan, the DIA argues in portentous defense-speak: "anti-government elements throughout the region undoubtedly would find a new generation of potential recruits."[13]

Lake Victoria may be the site of the first international "chronic emergency" caused by a bioinvasion. The governments of the three lake states, in conjunction with the World Bank, the Global Environment Facility (a World Bank–U.N. consortium), and various other international agencies, have launched an ambitious program that aims at comprehensive management of the lake's environmental crisis. About $78 million has thus far been pledged to the Lake Victoria Environmental Management Program. Hyacinth control figures large, and Aquatics Unlimited, a California-based company that has fought water hyacinth all over the world's tropical and warm temperate regions, has engaged the weed in Lake Victoria.[14]

But the future of the program is uncertain. Relations between Uganda and Kenya are tense, and Ugandan officials, citing potential environmental damage, have refused to let Aquatics attack the weed with herbicide. Aquatics argues that at this stage herbicide is essential for cutting the infestation down to a size that might make it susceptible to "biological control" in the near term. "Biological control," in this case, would involve the release of insects that eat the plant in its native range. (See Chapter 9.)[15]

Even if the hyacinth is brought under control, the problem of managing the other monster remains—a problem complicated immensely by human dependence upon it. But which humans? The old tilapia fisheries were largely artisanal subsistence activities: people fished for their own families and for local markets. The perch fishery is largely an export industry: most of the perch ends up in factories that process it, then ship it abroad. In general, fish is still the cheapest source of animal protein in the region, but demand from the factories more than doubled the price of perch from 1990 to 1996, and future scarcity is likely to push the price up far higher. Many people who once fished for themselves are reduced to buying scraps from the processors.[16]

Meanwhile, in the lake, the monster is not satisfied with sardines and prawns; it is now eating itself. Little perch are now important prey for big perch. In Tanzania, according to Tijs Goldschmidt, the perch's appetite for itself profoundly disgusts people; they say they fear that eating the fish could spread cannibalism among humans.

There is a kind of mythic truth in that response. A degraded landscape does somehow infect the societies that occupy it, and that makes the process of healing much more complex than it might otherwise be. What sort of recovery is possible in a "denatured" landscape—an ecosystem deprived of some of its basic components?[17]

★   ★   ★   ★

Lake Victoria's troubles have a precedent half a world away. Consider the ecological rebound of the North

American Great Lakes. Collectively covering 24.5 million hectares, these five lakes are the largest in the world—3.5 times the size of Lake Victoria. They carry more shipping than any other freshwater system on Earth, and their shores have seen some of the continent's heaviest industrial and agricultural development. Yet the lakes support a set of fisheries worth, at a very rough estimate, $4 billion annually. (This value is not a direct measure of the catch, however, since about $3 billion comes from sport fishing, which includes recreational revenues far exceeding the nominal value of the fish).[18]

By some measures, hard times in the Great Lakes have been worse than anything Lake Victoria has yet faced. In the early nineteenth century, lake stocks of Atlantic salmon, lake trout, whitefish, burbot, and other native fish constituted some of the greatest freshwater fisheries on Earth. Even allowing for the "fish-story" factor, the fishing must have been prodigious by modern standards—hooking 100 bass and walleye in a couple of hours was reportedly common. But by the beginning of this century most of the commercially important stocks were in decline. Mill dams had blocked access to spawning grounds in tributary rivers; the water itself was fouled with tannery waste, sewage, slaughterhouse offal, and silt from plowing and logging. The putrid inshore waters nursed typhoid, which erupted periodically into epidemics. Offshore, a rapacious fishing effort eventually rode the remnant stocks into oblivion.[19]

By 1960, every major fishery had collapsed. The lakes had become poisonous sores in a highly dysfunctional landscape: vast algal blooms on the water surface, anoxic dead zones below, surviving wildlife sickened from the DDT and other toxins that laced the water. The defining moment of the era arrived on June 22, 1969, in Cleveland, Ohio, on the shores or Lake Erie, when the brown and greasy Cuyahoga River, slick with oil and burbling methane from the rotting garbage in its depths, burst into flames.[20]

The Great Lakes also have their own version of the water hyacinth problem. There is, first of all, the sea lamprey, a kind of fish vampire. With its mouth clamped to the side of its victim, the parasitic, eel-like lamprey sucks blood and liquefied tissue from the wound that it makes with its rasplike tongue. It will move from one fish to another, eventually growing to more than 80 centimeters in length. The lamprey is native to the Atlantic, probably to the Saint Lawrence estuary, and possibly even to Lake Ontario, the easternmost of the Lakes, from which the Saint Lawrence flows. But no lamprey could have leapt Niagara Falls, which thunders into the west end of Lake Ontario from Lake Erie above. In 1921, the reengineering of the Welland Canal, a shipping detour around the falls, probably opened the upper four lakes to the lamprey. All the lakes were colonized by 1946.[21]

It was the large predatory fish, like lake trout and Atlantic salmon, that were preferred by both people and lampreys. Once these had disappeared, the lakes disintegrated into a kind of ecological free-for-all. Populations of smaller fish—both native and exotic— exploded and then collapsed once they ran out of food. The most spectacular of these events involved the alewife, another Atlantic native that had been skulking around the lakes and their canals for most of the century. Once free of the predators, it pushed the native white fish, bloater, and yellow perch into collapse by outcompeting them for plankton. During the late 1960s, the alewife periodically emptied its larder, and enormous alewife die-offs choked the shoreline.[22]

In 1955, the United States and Canada set up the Great Lakes Fisheries Commission to regulate the fisheries in a comprehensive fashion; killing the lamprey was a part of the organization's inaugural mandate. But despite an entire bureaucracy devoted to its destruction—and spending, at present, more than $12 million a year to that end—the lamprey is not going to let go of the lakes. Its population is being suppressed largely by a couple of chemicals that kill it in its larval form. But these "lampricides" share the disadvantages of other pesticides. They are so toxic that they must be carefully targeted, which means they can never reach all the lampreys. There is also, presumably, some danger of building a tolerance, although increasingly sophisticated targeting is reducing the amount of lampricides used, and therefore the likelihood of adaptation.[23]

Other techniques against the lamprey are beginning to ease lampricide use as well: special barriers designed to allow the passage of other fish but to trap lampreys, and "sterile male" releases that flood lamprey spawning runs with thousands of sterilized males, thereby diverting many females into unproductive unions. Even so, the uncertain relationship between lamprey and lampricide is now basic to the ecological balance of the lakes.[24]

So the lamprey is a source of continual anxiety. Here is Robert Davis, a former member of Congress from the lakeside state of Michigan, venting his angst at a congressional hearing in 1991: "Interestingly enough, you know, many critters that we have in this country whether they are spiders or whatever serve some useful purpose. The sea lamprey serves no useful purpose whatsoever. It is a villain. It is a villain that must be destroyed."[25]

The Great Lakes are full of such villains. A growing number of uninvited fish, mollusks, plants, plankton, and assorted other creatures has turned the lakes into a veritable soup of exotics. At least 141 exotic organisms have established themselves in the lakes or on their shoreline, and the current rate of invasion is estimated at one new organism annually. (See Table 4–1.) The most famous recent arrival is the zebra mussel, a little shellfish about the size of a kidney bean. Originally from the Caspian Sea region, the mussel made its North American debut in Lake St. Clair (between Lakes Huron and Erie), where it was probably released around 1986 from a ship's ballast water tank. (See Chapter 7.) The zebra mussel and a close relative, the quagga mussel, have since spread throughout the lakes

Table 4–1. Establishment of Exotics in the Great Lakes, 1810–1997

Years	Fishes	Invertebrates	Disease Pathogens Parasites	Algae	Plants	Total
		(number)				
1810–49	1				9	10
1850–99	6	4			23	33
1900–49	7	8	1	6	18	40
1950–97	12	17	2	18	9	58
Total	26	29	3	24	59	141

SOURCE: Edward L. Mills, Spencer R. Hall, and Nijole K. Pauliukonis, "Exotic Species in the Laurentian Great Lakes: From Science to Policy," *Great Lakes Research Review*, February 1998.

and many other waterways of eastern North America, largely by hitchhiking on boating equipment. Ultimately, the mussel may infest two thirds of the United States and much of southern Canada.[26]

Zebra mussels multiply at an incredible rate—a single female may release more than 5 million eggs in the course of a year. The mussels will completely encrust just about any available surface; when they run out of room, they encrust each other. Millions of dollars have been spent blasting them out of intake pipes with high-pressure hoses and scraping them off boats. (See Chapter 8.) They are imposing a huge ecological cost as well. In many waterways, zebra mussels are killing off the native mussels by encrusting their shells so heavily that they cannot open them to feed or breathe. North America has the most diverse freshwater mussel fauna in the world—nearly 300 distinct species and subspecies. In the Mississippi River basin alone, zebra mussels could drive as many as 140 native mussel species into extinction.[27]

But the zebra mussels' principal threat is their efficiency in stripping the algae that they eat out of the water column. A single adult mussel can filter up to a liter and a half of water in a day. In heavily infested parts of the Great Lakes, the water has an uncanny crystalline clarity: the algae have virtually disappeared, and so have the zooplankton—the tiny animals that feed on algae. Most of the lakes' major fish species depend on zooplankton and algae when they are young. It is not yet clear what the mussels are doing to the fish, but it is probably no coincidence that in Lake Erie, the most heavily fished lake, the value of the catch dropped from $600 million before the invasion to $200 million by the early 1990s.[28]

The mussels may also be resurrecting the lakes' burden of contaminants. As they filter debris out of the water, the mussels ingest the organochlorine pesticides, PCBs, dioxins, and other potent toxins that would probably otherwise have slipped through the food web and been buried in sediment. In Lake St. Clair, half the toxic burden is now believed to be inside zebra mussels—but it won't stay there. Since the mussels are now channeling so much of the food web through themselves, the chemicals are likely to seep into the whole community, beginning with the bottom-feeding fish, wildfowl, crayfish, and other organisms that eat the mussels or their excreta. This zebra mussel effect may explain why, after nearly two decades of progress, the levels of PCBs are once again rising in several of the region's top predators—in some lake trout populations, for instance, and in the eggs of the bald eagle.[29]

So the zebra mussel is at least as big a villain, to use Robert Davis's term, as the lamprey. The mussel, the lamprey, and most of the lakes' other uninvited exotics serve "no useful purpose," and no doubt officials would be delighted to destroy them if they could. But there is another type of villain abroad in the lakes—a band of useful villains that are in some ways analogous to the Nile perch.

As with Lake Victoria, the stocking of exotic fish was a sort of reflex response to the decline of the native fisheries. In the Great Lakes, exotic carp and salmonids (the family that contains trout and salmon) were being introduced as early as the 1870s. But it was during the 1960s, as the lampricides came into use, that the stocking effort became an industry in itself. Millions of fish are still sown every year—an effort that was costing $60 million in 1991, when the value of the stocking was last assessed. Restoring the native lake trout, once the dominant predator of most of the lakes, is one goal of the stocking program, but much of the effort has gone to bolstering the exotic salmonids, especially several Pacific salmon. As with Lake Victoria, the Great Lakes became almost entirely an exotic fishery. In 1900, 82 percent of the Great Lakes catch consisted of native salmonids; by 1966, native species made up only 0.2 percent of the catch. Of the 11 native fish species that once accounted for the bulk of the catch, 4 are now extinct and the others are at risk.[30]

The stocking operation does not seem to be yielding any sort of stable replacement for what has been lost. Outside of Lake Superior, there appear to be few self-sustaining lake trout populations, and the exotic salmonids are much more popular with fishers than they are with ecologists. The exotics prey on natives (including the native salmonids), they have brought in exotic diseases and parasites, and they have interbred with native salmonids. They have also, in effect, justified the alewife's presence; the alewife preys on lake trout fry but it is a main item on the exotic salmonids' menu.[31]

The Great Lakes are no longer some sort of aquatic version of a painting by Hieronymous Bosch, the fifteenth-century Dutch artist whose surreal landscapes are crawling with deformity. But they are not exactly of the Hudson River school either. They have recovered from a case of acute poisoning, but they appear to be in a state of chronic flux—the result of a combination of chemical and biological pollution. For the time being, their prescription calls for continuing doses of lampricide and regular stocking. But we don't understand the patient's condition well enough to know whether that therapy will work indefinitely.

★   ★   ★   ★

One of the biggest differences between chemical and biological pollution is that much of the latter is intentional. The accidental invasions—the biotic "oil spills"—are clearly an enormous and growing problem, but at least creatures like the zebra mussel or the lamprey have no constituency—no industries that profit from them, no public that values them. (For more on biotic spills, see Chapter 7.) It's a very different matter for the exotic salmonids and Nile perches of the world.

In fisheries, as in forestry and rangeland management, exotics still exercise a kind of quasi-magical allure. They offer the prospect of escape—from natural limitations, and often from the obligations of responsible management. Surely there must be some exotic that will make those waters productive, no matter how overfished, polluted, drained, or otherwise degraded they are. And so it is that most of the world's freshwater ecosystems—and a good many of its coastal ecosystems too—now harbor a local version of Lake Victoria's Nile Perch or the Great Lakes' exotic salmonids. The effects of these introductions can be very difficult to evaluate, but the evidence is mounting against many of these "useful villains."[32]

The Atlantic salmon, for instance, may soon have its revenge on Pacific salmon for what the latter are doing in the Great Lakes. Beginning in the 1980s, both the United States and Canada have permitted West Coast fish farmers to stock Atlantic salmon in ocean netpens—a form of aquaculture to which Pacific salmon are not well suited. Escapes from these netpens are common—storms rip into them and so do hungry sea lions. Atlantic salmon are not yet known to be spawning in the Pacific Northwest, but there is a good possibility that they eventually will. If they start to, they are likely to inflict the kinds of damage that often results when an exotic salmon invades the range of a native cousin: displacement through competition, loss of genetic integrity through interbreeding, and the spread of diseases.[33]

The widespread "ocean ranching" of various salmon species is likely affecting native fish stocks elsewhere too. Both the Atlantic and the Pacific species have been introduced into New Zealand, Australia, and Chile, for example. (None of these countries has native salmon.) Chile has become the world's second largest salmon producer (after Norway), but small-scale Chilean fishers say the salmon have decimated the native fish stocks on which the local industry had depended.[34]

Even the mass production of native salmon has caused serious deterioration of wild stocks. Hatchery-bred Pacific salmon are interbreeding with wild strains in the Pacific Northwest, or simply outcompeting them. In Norway, huge escapes of farmed stock have overwhelmed the gene pools of some wild strains of Atlantic salmon—and essentially eliminated them. In the worst incident, some 1.2 million fish departed the netpens during the winter of 1988–89. Today, some Norwegian rivers contain five times as many farmed salmon as wild ones. The United Kingdom may allow fish farmers to raise a form of Atlantic salmon that has been genetically engineered for fast growth—another potential avenue for genetic invasion.[35]

The rainbow trout is on a career path very similar to that of its close relative, the Pacific salmon. Like the salmon, the trout is native to western North America and like them, it has been stocked worldwide as an aquaculture and sport fish. It is now established throughout the temperate zones and even in some high-elevation tropical waters. In various places around the world—from South African streams to Siberia's Lake Baikal—it is being blamed for outcompeting or eating native fish, and its aggressive foraging is almost certainly working other changes in aquatic food webs. (In South Africa, for instance, it has nearly eliminated at least one insect species: an ancient rarity called the Gondwana relict damselfly.)[36]

A broader concern is the trout's role in transmitting a disease called furunculosis. In Europe, furunculosis has suppressed the native brown trout in the face of the rainbow invasion. Like the rainbow trout itself, furunculosis now has a worldwide distribution and is a chronic problem in salmonid culture facilities. In Australia, cultured rainbow trout are carrying another disease, a newly discovered virus of unknown origin, that is also extremely dangerous to several native fish species.[37]

Even North America has its share of rainbow invasions. In many western streams where it is not native but has been introduced, the rainbow has hybridized with its rarer relative, the cutthroat trout, thereby obliterating many cutthroat races. U.S. fisheries authorities have grown increasing wary of introductions that invite this kind of hybridizing. But many sport fishers have not: the movement of salmonids and other popular sport fish is a common kind of freelance biological vandalism. The single greatest threat to western North America's native freshwater fish is now thought to be exotics—many of them sport fish.[38]

Europe's brown trout has also been introduced all over the world for sport fishing, with much the same effect as the rainbow. In Australia and New Zealand, brown trout are suppressing native fish and inverte-brates. And in North America, the European trout has exacted a kind of revenge on the rainbow—in the form of one of its native parasites, the protozoan that causes "whirling disease." The disease gets its name from the fact that it makes its victims chase their tails. It was introduced in the 1950s but lay dormant for several decades. In the 1990s, for reasons that remain unclear, it awakened and began wiping out strains of rainbow and cutthroat trout. The brown trout is often hailed as one of the most successful fish introductions in the United States. Because it helps suppress the brown trout's competitors, whirling disease may help explain the trout's success. Among the native fish that are los-ing out to the brown trout is the "state fish" of California, the golden trout. The appetite of exotic trout is also endangering some of California's rarer frog species, such as the mountain yellow-legged frog.[39]

The Mozambique tilapia was the aquaculture fad fish of the 1960s for the tropics and warm temperate regions. Its supporters saw it as a sort of aquatic chick-en; its adaptability and quick maturation seemed to offer tremendous potential to protein-hungry develop-ing countries as well as to astute aquaculturists. The species was introduced into virtually every country with waters warm enough to support it. But as is so often the case, the same qualities that aquaculturists admire in it also made it a formidable invader. In Australia, India, the Philippines, and many other places, it is displacing food fish more valuable than it is, and rare endemic species too (that is, species with a single, limited native range). "A real nightmare," is how it looked to one fisheries authority in the Philippines, for example, where it has suppressed milkfish farming.

Even among aquaculture's least discriminating practitioners—officials still happily dumping their latest discoveries into lakes and streams—the Mozambique tilapia is now often conceded to have been a mistake.[40]

The aquatic chicken award has since been transferred to the Nile tilapia, which is now a major part of aquaculture's development machinery. Given the proven ability of other tilapia species to invade and displace native fishes, it is probably only a matter of time—and a fairly short time—before the Nile tilapia is found to be serious threat to aquatic biodiversity. Meanwhile, other tilapia continue to spread. In California and Texas, introduced tilapia may be adapting to cooler waters, which could allow them to move north. Off the coasts of Hawaii and some south Pacific islands, the Mozambique tilapia has made the jump to full-strength seawater.[41]

There are hundreds of lesser "useful villains" as well. An Asian catfish brought into the Philippines for farming has suppressed a native cousin—which happens also to be a better food fish. North American bass have suppressed native fish in Japan, South Africa, and Guatemala. Common carp and tilapia species have apparently pushed 10–15 of the Mekong River's native fish to the brink of extinction. The North American pike drove three fish endemic to a Turkish lake into extinction. Introduced whitefish, from continental Europe, have eliminated populations of Arctic char in Scandinavian lakes. The Chinese grass carp, introduced all over the world for aquatic weed control, has infected fish in Europe and North America with an Asian tapeworm. And in North America, the now ubiquitous common carp is a swimming reservoir of contagion: it has been found to harbor 138 exotic parasites—algae, fungi, protozoans, leeches, flatworms, tapeworms, and crustaceans.[42]

Nor are all the culprits fish. Oyster culture is a global conduit for mollusk diseases, some of which have spread from commercial beds out into natural oyster populations. The shrimp industry is moving shrimp pathogens all over the world. (See Chapter 8.) South America's golden apple snail, a failed food introduction throughout much of East and Southeast Asia, has become one of the region's worst pests on rice. And the bullfrog, a large, aggressive frog from the eastern and central United States, has been introduced here and there into western North America, Central and South America, southern Europe, and East Asia to produce frog's legs mainly for the restaurant market. In many of its new ranges, it appears to be eating the native frogs out of existence.[43]

As can be seen from the "rap sheets" of the villains just described, several motives are driving these managed invasions. The big, handsome species are often introduced for sport, while the tougher species—tilapia and carp, for example—are commonly used to create an artificial fauna for rivers rendered too dysfunctional by dams or pollution to support their native species. (Carp are so tough they can even survive pesticide spills.) Stocking reservoirs with exotics is standard practice virtually everywhere in the world; the "tailwaters" below dams are often stocked as well. Except for the upper reaches of large tropical rivers and the waters of the far North, it would now be hard to find a major lake or river anywhere on the face of the planet that does not contain exotics.[44]

But aquaculture is by far the largest force driving intentional aquatic introductions today. Aquaculture yields some 23 million tons of fish a year, or about 20 percent of the world's global harvest of fish. (See Figure 4–1.) That percentage is likely to grow, since all of the world's major wild fish stocks are overexploited and their yields have leveled off. Increasingly, governments and international agencies look to aquaculture as both a development tool and a means of supplying protein to an ever-growing human population, now pushing 6 billion.[45]

Yet this "Blue Revolution," as it is sometimes called, is a much murkier affair than its ideological precursor, the Green Revolution. For one thing, so much of the blue depends on the green: aquaculture uses huge amounts of feedgrain, and it may often be more efficient to divert that production for direct human consumption. For another thing, aquaculture's "crops" do

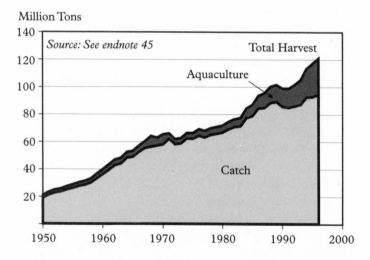

FIGURE 4–1. *World Fish Harvest, Fish Catch, and Aquaculture Production, 1950–96*

not generally stay put. A good deal of aquaculture is done in natural waterways; even where artificial ponds are used, escape is so common that the ponds may best be considered just another means of introduction. The consequent biological pollution is enormous, but the industry's willingness to reckon with it has thus far been limited.

Walter Courtenay, a biologist at Florida Atlantic University and an authority on fish introductions, defines the standard response: "those who introduce fishes purposefully are more interested in impacts on fishes considered immediately useful to humans and not to the overall native fish fauna or ecosystem; that is, if it cannot be caught and used, it is of little or no importance....whatever 'impact' studies they might conduct have the same narrow focus as the rationale for making an introduction." As presently practiced, aquaculture offers a substantial short-term payoff—but we are assuming long-term ecological risks that we cannot yet even calculate.[46]

5

Islands

For millions of years, thundering Pele has spewed forth her archipelago in the farthest reaches of the Pacific— farther from continental shores than all but a few scattered humps of land. Each of her islands breaks the waves black and steaming; each rises high above the water, then gradually erodes and sinks as newer islands explode from the sea. During the 5–10 million years that any one of these nodes spends above the waves, Pele tolerates a few terrestrial pilgrims. These arrive in a slow and intermittent procession. Sea birds with snails glued to their feet. Every couple of millennia, perhaps, a few forest birds blown off some distant island, or a raft of broken vegetation—typhoon flotsam from beyond the horizon, bearing an assortment of spiders, insects, and fertile fruit. Most of these arrivals die. But a few succeed, and over the eons Pele molds them into new forms: an island in evolution to match the islands in the sea.[1]

Pele may be a grudging host of mainland life, but the mainland, it turns out, loves her. Every year now, some 7 million tourists visit her realm. Yet the islands that they seek are intact only in a geological sense. Hawaii as a distinct ecological reality is rapidly vanishing.[2]

Hawaii's extreme isolation has acted as a kind of biotic filter: only certain kinds of creatures can survive such a crossing. Mammals, for instance, tend to have high metabolic rates: any monkey or tree shrew unfortunate enough to find itself on a log bound for Hawaii would have died of thirst or hunger long before its craft washed ashore. (Hawaii has only one native terrestrial mammal—a bat, which presumably flew rather than floated in.) Even the generally much tougher lizards and snakes never made it. (Hawaii's only native reptiles are sea turtles.) But any creature that did make it— assuming it arrived in sufficient quantity to establish a breeding population—would have enjoyed an enormous benefit over its mainland cousins. The colonists found relatively few competitors or predators.[3]

This easing of ecological pressure permitted evolutionary developments that the more crowded mainland ecosystems probably would have snuffed out. The islands produced more than 100 species of lobelia— tropical plants famous for their brilliant, trumpet-shaped flowers. Most lobelias are of rather modest proportions, but on Hawaii they were shrubs, vines, even small trees. There were about 50 species of honeycreeper—small, brilliantly colored, forest-dwelling birds. Some honeycreepers had an extravagantly long curved beak that matched exactly the extravagantly long curved flowers of the lobelias they pollinated while sipping nectar. The islands were full of this kind of evolutionary exuberance. There were more than 800 species of terrestrial snail—the greatest diversity of these creatures anywhere in the world. There were tree ferns that grew 8 meters high and produced fronds 3 meters long; there were 2-meter-high violets.[4]

In terms of its original biota, the Hawaiian archipelago is island ecology taken to some higher power. Compared with mainland ecosystems, most islands are relatively species-poor, but they usually contain a high proportion of endemics (species that live nowhere else). This tendency is strongest on "oceanic" islands (those lying off continental shelves), and on the Hawaiian archipelago it reaches its extreme. According to the most exhaustive inventory to date, about 17,600 species are thought to be native to the islands, and at least 9,000 or 10,000 of them—well over half—are endemic. Of course, the inventory includes thousands of fungi, protozoa, and other tiny creatures that have not yet received much scientific attention. Larger, better-studied creatures suggest that the real rate of endemism may be far higher: 91 percent of all native flowering plants are endemic, for instance; so are all six of the native freshwater fish, 99 percent of the mollusks (the snails and their kin), and 98 percent of native insects.[5]

The Hawaiian biota is distinctive in another way as well: it is going extinct at a rate that is breathtaking even by contemporary standards. During the past two centuries, at least 263 of the islands' native creatures have disappeared and another 360 or so are currently listed as endangered or threatened under the U.S. Endangered Species Act. (These numbers are indefinite because the classification of some organisms is uncertain.) Biologists have identified yet another batch of roughly 775 creatures that probably should be "listed." About 1,400 native Hawaiian life forms are therefore in trouble or already extinct, and for the most part these are creatures that command attention: birds, plants, and colorful snails.

Take the islands' birds. We know from current surveys and from prehistoric remains that the islands had at least 111 native birds (excluding the 168 "visitor species"—mostly seabirds that come to Hawaii regularly but spend little time there). Of those 111, 51 are extinct and another 40 are in serious trouble. Of the islands' 1,126 known types of native flowering plants, 93 are already extinct and another 655 are in trouble. Hawaii accounts for only 0.2 percent of the land area of the United States, but is home to 38 percent of the country's listed plants and 41 percent of its listed birds. The islands have been called the "extinction capital" of the United States.[6]

It is true that Hawaii has suffered a great deal from agriculture and logging—eucalypt and melaleuca were favored by island foresters for a time. But today bioinvasion in its purest form is the overwhelming threat to what remains of Hawaiian nature. Exotic predators, competitors, or diseases are a major pressure on more than 95 percent of Hawaii's endangered species. Think again about the plight of the native birds. Hawaii's first human colonists imported pigs, dogs, and the Polynesian rat; all these creatures prey on eggs or birds. The brown and black rats arrived with later colonists. As on many tropical islands, the small Indian mongoose was imported in a vain attempt to control the rats. The common barn owl, native to Eurasia and

North America, was brought in for the same reason. Rats, mongooses, and owls all attack native birds too. So does the common myna bird, a pet species gone wild. So does the domestic cat.[7]

But exotic predators are not the birds' only problem. Sometime in the 1820s, a crew of sailors came ashore to clean and refill the mosquito-infested water casks from their ship. The islands, which until then had no mosquitoes, were soon enveloped with them. Among the creatures the mosquitoes fed on were various exotic birds introduced by immigrants longing for more familiar avian life. (See Chapter 6.) From these birds, the mosquitoes picked up a couple of exotic diseases— avian pox and avian malaria—which they then transmitted to Hawaii's native birds. The native birds proved extremely susceptible; today, the diseases continue to force their numbers down. During recent epidemics, dying birds have literally just dropped out of trees. Elsewhere on the islands, aggressive exotic plants are destroying seabird nesting habitat. And pretty well throughout the archipelago, there has been a brute displacement of the native bird fauna by all those exotic birds. At least 46 species of exotic birds are now established on the islands and another 150 have been spotted but may not yet be breeding there.[8]

It is the same story with many of the islands' other creatures. Hawaii's extraordinary snails, for example, are falling prey to the rosy wolfsnail, a native of central America and the southeastern United States. The wolfsnail attacks and eats other snails—sometimes it eats them shell and all. It was released onto the islands in 1955 in the hope that it would control the African tree snail, an exotic crop pest. Naturally, it did not confine its attentions to the African snail; it has apparently driven dozens of native snail species into extinction. The rosy wolfsnail has been introduced onto more than 20 other island groups in the Pacific and Indian Oceans, with similar results.[9]

The native plants may have it worst of all. More than 100,000 wild pigs are rototilling their way through the islands' remnant rainforests, chewing up the under-brush, uprooting tree ferns and shrubs, and distribut-ing the seeds of exotic plants in their dung. Drier areas are visited by the other doomsday herbivore of island ecosystems: the goat. Bands of wild goats are stripping hillsides down to bare earth, which then erodes out to subsoil. Large tracts of dry forest were bulldozed in the 1970s to provide more habitat for exotic game birds. Surviving tracts are losing out to fire-adapted exotic grasses. (See Chapter 2.) Forests of all types are suffer-ing from the loss of native pollinators, both birds and insects. Even in the islands' healthiest surviving forests, most of the insect life is now exotic, and the exotic insects will not generally pollinate the native plants. One exotic that has penetrated lowland rainforest, the long-legged ant, is actually eating the tree pollen.[10]

Finally, as with the birds, the native plants are increas-ingly falling victim to brute displacement. A dense, prickly shrub called Koster's curse, introduced from the American tropics for erosion control, now infests more than 40,000 hectares on the island of Oahu alone. The fire tree, a nineteenth-century import from the Azores and Canaries, now infests all the major islands, covering more than 34,000 hectares. And the unstoppable banana poka, an Andean vine imported as an ornamen-tal, is green death to Hawaiian forests. With the aid of the pigs' digestive systems, it can penetrate virtually any plant community, where it will suffocate entire forests, from the canopy to the ground, in a thick, tangled mat that sometimes topples mature trees. In all, at least 877 exotic plants are currently invading Hawaii.[11]

Once the cogs start falling out of the machine, the whole contraption is likely to come apart. Lose a honeycreeper and you are likely to lose a lobelia, or vice versa. And the pressure on what is left of the ecosystem has grown fantastically intense. It is estimated, for example, that before people came to the islands, one new insect or other invertebrate probably established itself every 50,000 to 100,000 years. Today, about 20 new invertebrate species colonize Hawaii *every year*— the rate of invasion has increased at least 1 million times.[12] Hawaii is becoming a scruffy collection of pan-tropical weeds—the trademark of wasteland from Mexico to Nigeria. It is empty landscapes chewed by goats into a tropical version of the tundra. It is birdsong familiar to the inhabitants of Pittsburgh or Calcutta. It is tramp mosquitoes and ants. Pele now presides over a kind of ecological junkyard—littered scraps of mainland life.

★   ★   ★   ★

What is happening on Hawaii is happening on hundreds of other islands all over the world. On the western Pacific island of Guam, for instance, the brown tree snake has eaten 12 of the island's 14 land bird species into extinction in the wild, and several lizard and bat species as well. Aggressive, mildly venomous, and nocturnal, the snake is an accomplished climber and can grow to 3 meters. It is native to the Papua New Guinea region, and probably arrived on Guam around 1950 as a stowaway on military equipment. Virtually all native vertebrate species have been suppressed or driven into extinction since the snake's arrival. The snake itself is in no danger of dying out because it has switched its attention to various exotic lizards and birds—including poultry—and these are unlikely to go the way of the natives.[13]

Scientists fear that without major improvements in air and ship traffic inspection, it is only a matter of time before the brown tree snake colonizes other Pacific islands and perhaps some continental areas as well. At Honolulu International airport in Hawaii, for example, the snake is occasionally discovered in the wheel wells or cargo bays of planes arriving from Guam—despite vigorous efforts on Guam to keep the snakes out of the planes.[14]

Far to the south, in French Polynesia, the miconia tree, from the American tropics, is smothering virtually all of the territory's major islands. Its beautiful foliage has earned miconia a place in tropical gardens all over the world. But the tree grows to 15 meters and casts dense shade that excludes other vegetation. Seedlings reach sexual maturity in only a few years, then produce millions of seeds of their own. On Tahiti, where it is known as "the green cancer," miconia has displaced more than two thirds of the island's native forest, and is threatening 25 percent of its native wildlife species.[15]

In New Zealand, roughly two thirds of the land surface is covered by exotic plants. Nearly 60 percent of the plant species growing in the country are now exotic, and new plant species are establishing themselves at the rate of four a month. New Zealand has its own forms of "green cancer." *Clematis vitalba,* for instance, is an innocuous ornamental vine in its native northern Europe, but in New Zealand it smothers stands of trees

more than 20 meters high. After the vegetation collaps-
es, clematis blankets the resulting tangle, sometimes
with mats meters thick, preventing any regeneration.[16]

But it is exotic animals, particularly mammals, that
are causing the most trouble in New Zealand. (Like
Hawaii, the country's only native mammals are bats.)
The Polynesian ancestors of the Maori brought in the
Polynesian rat. Among the 80 or so vertebrates that
Europeans added were 32 mammals—the Norway and
black rats, weasels, cats, deer, and all the usual barn-
yard fauna of Europe. One of the most unusual—and
most regretted—additions came not from Eurasia but
from Australia. A little marsupial called the brush-
tailed possum was introduced into the country during
the nineteenth century to establish a fur industry, but
the possum developed a monstrous appetite for the
island's remaining native forests. Today, every New
Zealand sunset brings out roughly 70 million of them,
and before dawn another 21 tons of trees—bark, buds,
berries, leaves, flowers, and all—have disappeared
down the possums' digestive tracts. Nor is it just the
trees that suffer: the possum also eats the eggs and
chicks of some native birds, and displaces others from
their nesting sites.[17]

At least 35 of New Zealand's native bird species have
gone extinct, along with several species of reptiles and
large flightless insects—most of them probably eaten
into oblivion by predatory mammals (including
humans). Perhaps as many as 1,000 native creatures
are now at risk, and invasion is the most serious general threat they face.[18]

On the other side of the Pacific, about 800 kilometers off the coast of Ecuador, the Galápagos islands also
have a virulent case of invasion disease. Of the islands'

11 native species of bats and rats (their total complement of native terrestrial mammals), 8 are now extinct,
6 of them because of exotic pressure. And the process
continues. Wild pigs are eating the eggs of the famous
giant tortoises, the endangered green sea turtle, and
several iguana species. On Pinzon Island, black rats kill
virtually every hatchling of the local race of giant tor-
toise. On some islands, goats have eliminated all the
seedlings of several native tree species, and wild house
cats have eaten most of the lava lizards. An exotic ant,
the little fire ant, has suppressed most of the islands'
native ant species.[19]

On Floreana Island in the Galápagos, thickets of the
exotic scrambling shrub *Lantana camara* are encroaching on the nesting grounds of the darkrumped petrel.
The guava, a South American fruit tree that infests
some 40,000 hectares of the islands, fueled a fire in
April 1994 that destroyed some 4,000 hectares of
woodland on Isabela, the largest island in the group.
And on several islands, a major bramble invasion has
apparently begun to take shape.[20]

On the Indian Ocean island of Mauritius, 900 kilometers west of Madagascar, invading plants now dominate the landscape so completely that the native flora
is limited to eight fenced-in plots, varying in size from
1.5 to 15 hectares, and a couple of tiny nearby islets.
The plots must be hand weeded to prevent them from
disappearing into the surrounding carpet of exotics—
731 invading plant species have been counted thus far.
The Mauritian fauna is just as exotic as the flora: pigs,
goats, deer, rats, monkeys, rabbits, and giant African
snails are among the more disruptive exotic animals.
The rosy wolfsnail is here too—and has driven 24 of
the island's 106 endemic snails into extinction.[21]

★   ★   ★   ★

In October 1835, Charles Darwin was exploring the Galápagos Islands, amusing himself with the local birds:

> There is not one which will not approach sufficiently near to be killed with a switch, and sometimes, as I have myself tried, with a cap or hat. A gun is here almost superfluous; for with the muzzle of one I pushed a hawk off the branch of a tree. One day a mocking-bird alighted on the edge of a pitcher (made of the shell of a tortoise), which I held in my hand whilst lying down. It began very quietly to sip the water, and allowed me to lift it with the vessel from the ground.[22]

Early visitors to remote islands often commented on the Edenic naiveté of the local animals. The Galápagos birds did not evolve in the presence of quick, smart mammalian predators, so they were not predisposed to view a giant, exotic naturalist as dangerous. Island plants are often "naive" as well. On the Hawaiian islands, for example, some plants have shed the thorns that armed their immigrant ancestors; in the absence of large herbivores, thorns are a waste of photosynthetic effort. Hawaii's tree ferns are so easily injured by pigs partly because they have very shallow roots: they did not evolve in the presence of anything that would want to push them over.[23]

This "naiveté" goes some way toward explaining why islands are not just centers of endemism, but centers of extinction. Exotics often overwhelm an island because there is no local version of them—nothing that would have "prepared" island creatures for the invasion. Nothing in their evolutionary past prepared the Galápagos tortoises for rats; nothing prepared Guam's birds for the brown tree snake. New Zealand's trees evolved without mammalian herbivores: they do not produce potent chemicals to discourage browsing, the way eucalypts do in Australia, where the possum comes from. On the nitrogen-poor volcanic slopes of Hawaii's big island, none of the native plants can "fix" nitrogen; they never had to compete with shrubs like fire tree, which can.[24]

The collision between mainland and island biotas had until the past couple of decades been the principal cause of historic-era extinctions—at least among the kinds of creatures we are good at observing. The trend is easiest to see in birds. Birds are ubiquitous and conspicuous; people generally like them and they notice when birds are no longer around. So it is possible to piece together a fairly comprehensive record of bird extinctions during the past four centuries, and the record shows that 91 percent of those extinctions occurred on islands. The autopsies on these events vary somewhat in their conclusions, but invasion is generally cited as the principal cause of the trend. (One survey, for example, attributes 57 percent of these extinctions to exotics; another ascribes 70 percent of them to exotic rats alone; still another concludes that mammal invasions of islands remain the most serious threat to the world's seabirds.)[25]

Mammals show a similar trend, but the mammal picture is a good deal murkier. Unlike birds, mammals are often difficult to find, and they are relatively rare on oceanic islands. Even so, 36 percent of known mammal extinctions since 1600 have been island affairs. In all, about 75 percent of historic-era vertebrate extinctions are thought to have occurred on islands. (See Table 5–1.)[26]

Extinction is a natural process. Like individual

Table 5–1. Historic-Era Terrestrial Mammal and Bird Extinctions

Area	Mammals	Birds
	(number)	
Continents		
Africa	11	0
Asia	11	6
Australia	22	0
Europe	7	0
North America	22	8
South America	0	2
Total	73	16
Islands		
Near continents	8	7
In Pacific Ocean	4	109
In Indian Ocean	4	18
In Atlantic Ocean	23	20
In Mediterranean Sea	2	1
Total	41	155

SOURCE: Robert L. Peters and Thomas E. Lovejoy, "Terrestrial Fauna," in B.L. Turner II, ed., *The Earth as Transformed by Human Action: Global and Regional Changes in the Biosphere over the Past 300 Years* (Cambridge, U.K.: Cambridge University Press and Clark University, 1990).

organisms, species have finite lives. No kind of living thing will live forever: eventually, it will either die out or evolve into something else. So in the very long run—in what scientists call "evolutionary time"—it is the rate of extinction rather than the fate of any particular species that will tell you the most about the fortunes of biodiversity. On a local level, that rate will vary greatly, of course, depending on community conditions and the type of organism concerned. But on a global level, the skein of life possesses an enormous and supple durability; a big ecosystem may last for millions of years, losing some species, gaining new ones, expanding and contracting as the climate changes. Such durability suggests that most of the time, extinction is a low-level "background" phenomenon; only occasionally does it become a monster that swallows up hundreds of species at once.

So what is the "background rate" of extinction? How long should a species last without human interference? Fossil evidence—even though it is vague and hard to interpret—offers the best hope of an answer. Surveys of marine invertebrate fossils, for instance, suggest that ancient species of this type had an average life of from 1 million to 10 million years. Fossil terrestrial vertebrates seem to have had a shorter average life-span, on the order of only 1 million years.[27]

And in the broadest terms, the fossil record offers another basic insight into global extinction rates. If you take the record as a global whole and look at it in huge, 10- or 100-million-year chunks, what you see is a general trend toward more species—toward greater diversity. The trend does not necessarily hold for particular classes of organisms or for finer time scales, but on that macroscopic level, it seems to have been interrupted only 11 times in the entire history of life. (The most famous of these interruptions occurred at the end of the Cretaceous Period, about 65 million years ago, when the dinosaurs went extinct.) Apart from those 11 major extinction events, the rate at which new species evolve appears generally to have been higher—and usually far higher—than the rate at which they have gone extinct. (See Table 5–2.)[28]

That is clearly not the case today. Current extinction rates for some of the best-known groups, like birds or mammals, are thought to be 100 to 1,000 times their

Table 5–2. Major Global Extinction Events

Period	Time Since Event	Share of Species Thought to Have Gone Extinct
	(million years)	(percent)
Middle Miocene	12	24
Late Eocene	35	35
End of the Cretaceous	65	76
Late Cenomanian	90	53
Aptian	116	41
End of the Jurassic	146	45
Pleinsbachian	187	53
Late Triassic	208	76
Late Permian	245	96
Late Devonian	367	82
Late Ordovician	439	85

SOURCE: Michael L. Rosenzweig, *Species Diversity in Space and Time* (Cambridge, U.K.: Cambridge University Press, 1995).

ancient "background rates." If those famous "endangered" and "threatened" categories turn out to be good predictors of future extinctions, then extinction rates could jump another 10-fold by the end of the next century. This twelfth great agony of extinction began on islands, but it has now infected the world's mainlands. Around 1930, the extinction count on the continents— at least when it comes to conspicuous animals—seems just about to have caught up with the island count. For invertebrates and perhaps also for plants, the island-mainland distinction has likely disappeared as well. For the past few decades, the primary cause of extinction has been the destruction of tropical forests, whether on islands (the Malay archipelago, for example) or continents (Amazonia).[29]

Tropical forests contain the world's richest ecosystems, in terms of the number of species inhabiting them. Most of this diversity consists of an overwhelming abundance of arthropod species (insects and their relatives), and it is mainly the estimated loss among these creatures that is swelling the extinction rolls today. No one can say for sure what the global rate of extinction is, but a conservative estimate might put it at 4,000 species per year, or about 11 per day. (The estimate is conservative because it uses two low-end numbers: it posits a global total of only 5 million living species and it assumes that only 50 percent of them inhabit tropical forests.) This "best case" scenario would put the current life span of an "average" species at something like 500 years—only 0.05 to 0.005 percent of the "natural" life spans cited earlier. On the scale of a 75-year human life span, that percentage would represent from two weeks to less than a day-and-a-half.[30]

Despite the now global extent of the extinction crisis, islands remain crucial for understanding it. In part that is because there are a lot of ecological "islands" on mainlands. A lake, for example, is a kind of island in reverse—an isolated lode of water in an ocean of land. And lake life is island life: think of Lake Victoria, with its flock of haplochromine cichlids (see Chapter 4), then think of Hawaii with its flocks of honeycreepers and snails. Both places are vessels of endemic diversity; both show the island trait of elaborate local variation on a single, ancestral theme.

There are wholly terrestrial "islands" too. At the southern tip of South Africa, in the highlands east of Cape Town, there is a 9-million-hectare expanse of shrubland growing on dry, nutrient-poor, sandy soils—

an upland archipelago above the richer, moister low-land soils, now largely converted to agriculture. This apparently unpromising region is the Cape Floral Kingdom, which is thought to contain the most diverse plant community on Earth—8,574 native plant species have been identified, and 68 percent of them are endemic. Like the cichlids and honeycreepers, the endemic Cape flora is largely an array of closely relat-ed flocks: 69 shrub species belong to the genus *Protea*, another 80 to *Leucadendron*, and so on. And here, too, the extinction disease is raging: 58 of the Kingdom's plant species are already gone; another 3,435 are threatened to one degree or another.[31]

And as with Lake Victoria and Hawaii, the primary agents of destruction are exotics. Exotic trees have dis-placed native vegetation over thousands of hectares. (See Chapter 3.) Less visible but perhaps more ominous is the arrival of the Argentine ant, which is dislodging the native ants. The Cape Floral Kingdom is in part a gar-den planted by its ants. About 1,000 of the native plants are known to rely on the local ants to bury their seeds—a service the Argentine ant refuses to perform.[32]

Once you start looking for these ecological islands, the continents seem to be speckled with them: alpine communities, lakes, desert springs, distinctive soil "islands" weathering out of unusual bedrock. These places, which are so important in the tapestry of conti-nental life—are they all as vulnerable to invasion as the geographic islands are? For many of them, we have as yet no way of knowing. But the fate of such places as the Cape Floral Kingdom suggests that wherever a rich lode of endemics lies, there lies a soft spot for the can-cer of invasion—a place where life gone haywire can gnaw away at itself. And because the endemic communities are so distinctive, their loss would be a far greater tragedy than their relatively small size might suggest. If the extinction crisis swallows up these islands of endemism, it will have done much greater damage to the planet's biodiversity than could be captured in a simple species count.

Islands are important for understanding the extinction crisis in another way as well: most of the Earth's remaining wildlands are becoming islands in a sea of developed or degraded landscape. This process can be observed in practically any terrestrial biome—wild prairie, for example, or wetland. But it is probably doing the most damage in forests, since forests may hold up to 90 percent of all terrestrial species. The world's original forest cover has been roughly halved, or reduced by nearly 3 billion hectares. About 60 percent of what remains consists of scattered patches— islands, in other words. Only about 20 percent of the Earth's surviving forest is in large, consolidated, and more or less natural ecosystems. The current rate of forest loss is somewhere around 14–16 million hectares per year; it may have dropped toward the lower end of that range since the 1980s, but it is still far higher than at any time in the historical past. From 1920 to 1950, for example, the average annual rate of deforestation was just 60 percent of the current rate. From 1700 to 1850—when the colonial era was in full swing—defor-estation probably averaged little more than 10 percent of what it does today.[33]

The fragmentation of forests and other natural areas is not absolute, since what isolates one species may not isolate another. An island of old growth forest sur-rounded by tree plantations, for example, would isolate creatures dependent on old growth, but not necessarily other forest animals. Even allowing for a degree of continuum, however, there are some important ecological reasons why the big outdoors will always die when it is cut up into pieces.

The basic dilemma can be expressed in a simple maxim: big habitats hold more species than little habitats. When you carve a big habitat up into little, isolated chunks, you do not get to keep the original set of species in each little chunk—some species will eventually die out in all the chunks. The victims of the process, at least initially, tend to be species that require a great deal of room (rhinos, for example) or species with quite specialized requirements that get harder and harder to satisfy (honeycreepers). Eventually, less demanding species tend to go as well, because smaller populations are more susceptible to disasters (natural or otherwise) and to the unhealthy effects of inbreeding.[34]

These processes help explain why virtually every census of fragmented tropical forest has found a loss of species. Even fragments specifically set aside for conservation are almost never large enough to escape the trend. A recent study of mammals in western North American parks, for instance, found that fragmentation had caused 29 population extinctions (that is, extinctions of a particular population, but not necessarily of an entire species). As time passes, and as the isolation of these places grows more absolute, they are likely to hemorrhage more species.[35]

"Habitat loss" is the bland standard label for the human fury that burns through a landscape and reduces its natural wealth to these remnant, pathological archipelagoes. And the habitat is almost always lost to an exotic biota serving human consumption or testifying to human waste. Crops, cattle, and goats. Waste-lot weeds and plantation trees. Aggressive grasses, tramp ants, and sparrows. So invasion pressure tends to grow as fragmentation proceeds, partly because the remnant wild areas are awash in a sea of exotics, and partly because simplified or otherwise disturbed ecosystems are generally more vulnerable to invasion.

Even in the fragments we call parks, invasion seems to be as inexorable as the direct effects of isolation. In the late 1980s, for example, a review of 24 field studies of natural areas in various parts of the world found exotics in all of them, except for some protected areas along the Antarctic coast. In continental U.S. nature reserves—among the most carefully scrutinized and protected pieces of real estate on the planet—anywhere from 5 to 25 percent of the flora is now exotic.[36]

As habitat dwindles, exotics are therefore likely to become increasingly dangerous, not just to the occupants of natural islands, but to the occupants of these "unnatural" fragments as well. In the United States, which has some of the world's most fragmented natural areas, the process is well advanced. Exotics are already the most common reason—although not usually the only reason—for listing an organism as officially protected under the U.S. Endangered Species Act: between 35 and 43 percent of all listed organisms are under pressure from exotics. Worldwide, according to the most comprehensive recent survey, nearly 20 percent of the endangered species in the world's best-studied animal groups—the vertebrates—are under pressure from exotics. (See Table 5–3.) And after habitat loss, invasion is now sometimes regarded as the second most important cause of endangerment on a global scale.[37]

But move away from the crucibles of extinction—the islands, whether natural or otherwise—and into the

Table 5–3. *Threatened Animal Species Known to be Under Pressure from Exotics*[1]

Species	Continental Species	Island Species
	(percent)	(percent)
Mammals	19	12
Birds	5	38
Reptiles	16	33
Amphibians	3	31
Freshwater fish	34	
Average	13	31

[1]The percentages are of total threatened species, not of total species within the taxonomic category. Figures are likely to be underestimates, since pressure from exotics can often be difficult to detect.

SOURCE: See endnote 37.

vast, indeterminate spaces that make up most of the world: in these places, endangerment may not be the best measure of ecological health. Many such spaces are badly invaded. (See Figure 5–1.) On the Great Plains of North America, for instance, between 30 and 60 percent of the plant species are exotic. But how sick are the plains or other such places because of that? "Think of it as a carrying capacity problem," notes Bruce Coblentz, an invasion biologist at Oregon State University. A patch of North American prairie will only be graced with a certain amount of sunlight and water: if cheat grass is now absorbing half of that, then the native prairie plants are losing ground, literally, and may suffer a serious loss of evolutionary potential, even though none of them may be in any immediate danger of extinction.[38]

Habitat loss and invasion: in a sense, these are two

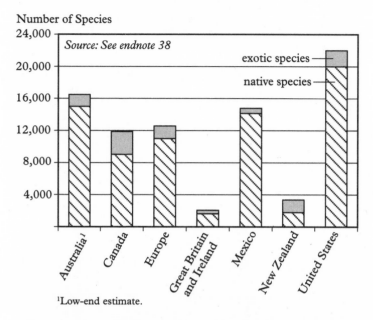

Number of Species

FIGURE 5–1. *Number of Exotic Plant Species in the Total Flora of Various Regions*

phases of a single disease. Intact habitat dissolves into islands amidst a sea of degraded landscape, then the islands dissolve too, as their native species die off and exotics invade. The result is impoverishment and monotony. It is not so much an empty landscape—a "silent spring," to use Rachel Carson's famous term—as a homogenized one. House sparrows sound the same, whether you are in Belgium or British Columbia or Hawaii.

In much of the world, the first phase of the disease has reached an advanced stage: relatively few large wild areas remain intact. The second phase is picking up momentum rapidly, according to Jeff Crooks, a biologist at the Scripps Institution of Oceanography, and Michael Soulé, the retired University of California ecologist who was a founder of the Society for Conservation Biology. In a recent paper, they argue that "it may not be long before invasive species surpass habitat loss and fragmentation as the major engines of ecological disintegration."[39]

II

The Culture of
Invasion

6

Colonists

A ship comes into New York Harbor in the spring of 1890. It's a passenger steamer, but some of the passengers disembark in cages: about 60 dark, voluble birds. Blackbirds, a contemporary New Yorker might have guessed. These are released in Central Park; a year later, they are joined by 40 or so of their fellows. How long did it take Europeans to colonize all of North America? Ponce de León claimed Florida for the Spanish crown in 1513, but the tall grass prairies of the Midwest were not entirely plowed out and grazed over until about a decade before that steamer came in. From the arrival of Ponce to the loss of the prairies, more than three-and-a-half centuries elapsed. *Sturnus vulgari*, the European starling, completed its conquest of the continent—spreading from Central Park to San

Francisco—in a little more than 50 years.[1]

The starling owes its New World debut to Eugene Schieffelin, the eccentric scion of a New York pharmaceuticals manufacturer. Schieffelin apparently had little interest in the family business. Portrait painting, genealogy, and church history were more to his taste, but his main goal in life seems to have been joining and founding societies. One of the societies Schieffelin founded was the American Acclimatization Society, which had as its aim "the introduction and acclimatization of such foreign varieties of the animal and vegetable kingdoms as may be useful or interesting." The Society released many other "useful or interesting" European birds besides the starling—thrushes, finches, skylarks, and nightingales. Schieffelin even played a bit part in bringing the house sparrow to the New World. But apart from the sparrow, none of these other releases succeeded. Schieffelin's only lasting gift to the continent is the starling.[2]

Schieffelin is often said to have wanted to establish in the New World every species of bird mentioned in the works of William Shakespeare. Apparently, no record of any such ambition on his part survives, although he is known to have founded a society called the Friends of Shakespeare. If the charge is true, however, Schieffelin fashioned a huge biotic drama out of a single line. There's just one reference to starlings in the Bard's entire works: "I'll have a starling taught to speak nothing but 'Mortimer'," says Hotspur, in Henry IV, part I, after the King forbids him to mention his brother-in-law, a suspected traitor. (Starlings are talented mimics.) There are now more than 200 million of them in North America, where they are a serious pressure on native cavity-nesting birds.[3]

★　　★　　★　　★

A fantastic and completely frivolous ambition, to make the New World sing in Shakespearean verse. But whether Schieffelin intended it or not, such an idea was perfectly in keeping with acclimatization societies like the one he founded. During the nineteenth century, a loose network of these societies, along with botanical gardens and fish hatcheries, moved hundreds of plant and animal species between North America, Europe, East Asia, and various European colonies. (See Table 6–1.) The aims of this network went far beyond a simple extension of the European farmyard biota. An interest in transplanting skylarks and starlings may seem absurd today, but it was the expression of a profound biological undertaking: the reconstitution of non-European landscapes as a whole.

Acclimatization was a later chapter in the history of European colonization, but it followed logically from the first pages of the volume. Even when these brave new worlds turned out to be prosperous, they seem often to have engendered a kind of colonial angst—an anxiety of difference. Today, for various reasons—the speed of jet travel, the spread of so many European species, the loss of so many indigenous cultures—that angst is difficult to reconstruct. But in the early phases of colonial expansion, these ends of the Earth, so far from Europe and so unlike it, could be frightening or repugnant to European sensibilities—even in features that later generations have come to celebrate.

New England, for example, is famous for its fall foliage, especially the incandescent red and orange blaze of its native maples. The maples are a source of regional pride and considerable regional income, from the tourists—the "leaf peepers"—who arrive to view

Table 6–1. Key Dates in Establishment of Selected Acclimatization Societies and Botanic Gardens

Year	Society or Botanic Garden
1543	Garden in Pisa, Italy
1652	Dutch East India Company's garden at Capetown, South Africa
1735	Pamplemousses botanic garden in Ile de France
1759	Royal Gardens at Kew, England, reorganized as botanic gardens
1774	Jamaica Botanic Garden
1787	English East India Company's Calcutta Botanical Garden
1796	Garden at Penang, Malaysia
1816	Sydney Botanic Garden, Australia
1817	Garden at Bogor, Java
1821	Garden at Peradeniya, Sri Lanka
1845	Royal Botanic Garden at Melbourne, Australia
1854	La société zoologique d'Acclimatation (Paris)
1858	Akklimatisations-verien (Berlin)
1859	Comité d'Acclimatation d l'Algérie (Algiers)
1859	Missouri Botanical Garden, United States
1860	Acclimatisation Society of the United Kingdom (London)
1861	Società di Acclimazione (Palermo, Sicily)
1861	Acclimatisation Society of Victoria (Melbourne)
1863	Nelson Acclimatisation Society (New Zealand)
1863	Botanic Garden at Christchurch, New Zealand
1864	Imperial Russian Society of the Acclimatisation of Animals and Plants (Moscow)
1871	American Acclimatization Society (New York)
1872	Arnold Arboretum, Massachusetts
1892	Botanical Garden of Buenos Aires, Argentina
1898	Entebbe Botanical Garden, Uganda
1930	Hui Manu, the Hawaiian acclimatization society

SOURCE: Christopher Lever, *They Dined on Eland: The Story of the Acclimatisation Societies* (London: Quiller Press, 1992); Quentin C.B. Cronk and Janice L. Fuller, *Plant Invaders: The Threat to Natural Ecosystems*, WWF and UNESCO "People and Plants" Conservation Manual 2 (London: Chapman and Hall, 1995); Stephen A. Spongberg, *A Reunion of Trees: the Discovery of Exotic Plants and Their Introduction into North American and European Landscapes* (Cambridge, MA: Harvard University Press, 1990).

the spectacle. In the 1640s, however, leaf peeping was not in vogue. In his *History of Plymouth Plantation,* William Bradford, the Puritan governor of Plymouth colony, observed sourly that fall imparts to this "hideous and desolate wilderness" a "wild and savage hue." (A reminder, that is, that New England is not England.)[4]

By the nineteenth century, these "neo-Europes" had largely ceased to be frightening, but in the eyes of the acclimatizers they were still degenerate, still requiring improvement in ways large and small. And the task was too important to be left to eccentric entrepreneurs like Schieffelin.

One of the most comprehensive efforts in this direction began in 1871, when the U.S. Congress set up its Fish and Fisheries Commission. Spencer Fullerton Baird, its first Commissioner, recognized an incipient crisis in the country's fisheries: Baird was one of a small company of naturalists who realized that many U.S. fish stocks were already in decline. Baird had grave doubts about whether the rapidly growing republic could continue to live from its fields, and he was convinced that "water can be made to yield a larger percentage of nutriment, acre for acre, than land"—a plausible supposition in the days before chemical fertilizers and high yield grains.[5]

So Baird set up an immense stocking program for fish of all kinds. During the next couple of decades, shad, salmon, trout, whitefish, bass, pike, haddock, and many other species were distributed by the commission. One of Baird's favorites was the common carp, from Europe. Carp, Baird thought, would give the southern United States an equivalent to trout; it would cost only half as much to produce as poultry, and carp ponds could be dug in land unsuitable for crops. The commission devised special fish cans for shipping baby carp to farmers all over the country. Baird even cajoled freight companies into shipping all the cans back for free.

Today, the carp is probably the most common freshwater fish in the United States. Similar programs elsewhere have made it a true cosmopolitan—occurring virtually everywhere there is available habitat, and growing fat in farm ponds and natural waterways all over the world. It is widely valued for its ability to tolerate a broad range of conditions, including heavy pollution. It is also widely condemned as a nuisance: the carp is the aquatic equivalent of a pig. It spends its time rooting around in the muck and muddying the water—a habit that lowers oxygen levels, interferes with sight-feeding fish, and releases nutrients that promote algal growth. And like a pig, it will eat just about anything, including any fish eggs that it can find. In some countries, including the United States, it is suppressing native fish.[6]

But Baird's ambitions ran well beyond conventional aquaculture. He managed to convince more than 100 railroad companies—virtually the whole of the country's rapidly growing and rather chaotic rail system—to take his fish cans aboard "and in many cases, to stop the car at stations near rivers or streams to allow the introduction of fish therein." By the early 1880s, a couple of specially adapted railroad "fish cars" were trundling through the countryside with more than 9 tons of cargo. During the next couple of decades, there were few, if any, watersheds in the country that did not feel the plunge of exotic fingerlings. And by the 1880s, the United States was exporting eggs of its native fish abroad—to Canada, much of western Europe, New

Zealand, Australia, and even the Sandwich Islands.[7]

It was in Australia that acclimatization approached its logical extreme. As in the United States, the activity was partly an attempt to improve the farm economy, although in Australia's case the immediate worry was too much labor rather than too little food. The problem was at its worst in the southeastern province of Victoria. Toward the end of the 1850s, Australia's gold rush had run its course, and hordes of unlucky miners were looking for work. In an essentially agrarian economy, the obvious answer to unemployment is more farms. The government passed a series of acts intended to encourage homesteading, but the land itself would also have to be made more productive. Couldn't that task be achieved quickly and cheaply by importing more productive plants and animals? During the 1860s, a half-dozen major acclimatization societies, fortified with both private and public funds, rose to the task. An assortment of fish, birds, mammals, and plants arrived from both the Old and New Worlds.[8]

As in the United States, acclimatization became a form of civic good works. Australian nature appeared to be badly flawed and the fortunes of the colony depended on its correction. If the eucalypts were riddled with wood-boring larvae that jeopardized their value as timber, then woodpeckers ought to be brought in to deal with the problem. Africa's snake-hunting secretary birds might be a good solution for all the snakes. And as for the local game—aborigines might content themselves with roast monitor lizard, but the landscape would have to be stocked according to the colonists' tastes, with rabbit, deer, and pheasant. And so by touches both broad (rabbits) and refined (songbirds and fireflies) the societies attempted to make over the strange nature of Australia.[9]

Behind the furious effort lies that colonial angst, apt to be far sharper in Australia than in North America. Australia is so obviously alien to the cheerful clutter of the idealized European countryside, where nature seems, by and large, a kind of furniture for humanity. Australian nature is definitely not furniture. The island continent, ancient and inscrutable, was not molded by any history a European could read, did not answer to people in any way a European could hear. The silence of the bush was suffocating. In 1857, a legislative committee endorsed the idea of wholesale acclimatization as a way of populating the solitudes, where "the almost unbroken repose of ages holds its sway." The project would involve the creation of a kind of selective and cosmopolitan Eden, as in this reverie from the newspaper *The Age,* of April 2, 1858, which hoped:

> to see the horse-chestnut and the oak add grandeur and variety to our woods, to have the Chinese sugarcane filling the cultivator's purse, to hear the nightingale singing in our moonlight as in that of Devonshire, to behold the salmon leaping in our streams as in those of Connemara or Athol, to have antelopes gladdening our plains as they do those of South Africa, and camels obviating for us as for the Arab the obstacle of the desert.[10]

Most of the introductions failed. The fireflies didn't take; neither did the peacocks, llamas, or ostriches. But the acclimatizers had a hand in several of the continent's worst invasions. European rabbits, for example, had been part of the colonial equipment since the arrival of the First Fleet, in 1788, but the societies were the main engines driving their establishment in the wild. The effort proved a disastrous success. For nearly a century, until the introduction of a Brazilian rabbit virus in the 1950s, billions of rabbits plagued Australia. (See Chapter 9.) Despite immense pogroms in which tens of millions of rabbits were killed, despite more than 3,200 kilometers of fencing designed to block their advance, the rabbits gnawed Australia's range into stubble and dust. Sheep and cattle starved; farmers abandoned their land.[11]

Among the other exotics that the acclimatizers helped spread were European starlings and sparrows, which suppress native bird populations, and various species of deer. They also introduced the prickly pear cactus, which went on a vegetable equivalent of the rabbit rampage, overrunning more than 310,000 square kilometers of rangeland, before it too was stopped by an introduced biocontrol agent, the moth *Cactoblastis cactorum*. (For a discussion of biocontrol, see Chapter 9.)[12]

*   *   *   *

Since acclimatization was in some measure a colonial enterprise, its enthusiasts often expected it to yield some sort of return for the mother country. In mid-nineteenth-century Algeria, for instance, French colonial authorities had hopes of covering much of the countryside with eucalypts and bamboo, in order to make the climate moist enough to grow a wider assortment of crops. After the Franco-Prussian war (1870–71), French acclimatizers hoped Algeria's introduced eucalypts would compensate for the lost forests of Alsace-Lorraine. But the biggest benefit was colonization itself. Reforming Algerian nature—and agriculture—on a European model would open up living space for French smallholders, who could occupy the landscape and make it profitable. Europeans themselves would therefore need to be "acclimatized" to their colonial environs—a notion that became commonplace in colonial ideology. Auguste Hardy, colonial Algeria's chief botanist and director of the formidable experimental gardens of Algiers, once wrote that "the whole of colonization is a vast deed of acclimatization."[13]

Despite their own need for improvement, it was possible that the colonies could contribute beneficial organisms to Europe. "We have given the sheep to Australia; why have we not taken the kangaroo—a most edible and productive creature?" asked Isidore Geoffroy Saint-Hilaire, the first president of the French Société Zoologique d'Acclimatation. "New conquests of animals and plants will serve as new sources of wealth." The Société became the preeminent organization of its type in Europe. It owed its founding largely to Geoffroy Saint-Hilaire's conviction that "all civilized countries" ought to collaborate on a project "to populate our fields, our forests and our streams with new inhabitants" as a means of increasing the productivity of the European landscape.[14]

The search for useful animals even inspired a new (and short-lived) branch of zoology. Geoffroy Saint-Hilaire coined a term for it: "zootechnie"—a kind of animal technology. In an era not yet dominated by engines and computers, it was conceivable that unusual species of animals could augment industry. Is a new beast of burden really any less plausible a token of progress than a new model of car? For a time, Burchell's zebra could be seen pulling wagons through the streets of Paris. Yet the societies' contributions to the European fauna proved fairly modest. They helped establish certain game birds, such as pheasants, and various North American salmonid fish. The English society may also have had a hand in the introduction of the east Asian sika deer, which has become a forest pest in Scotland and is interbreeding with the native red deer.[15]

Plants, however, are a very different story. The European powers began to establish a network of botanical gardens very early in the colonial period. There was a Dutch garden in Capetown, South Africa, by 1694, a French garden on Mauritius by 1735, and English gardens in Jamaica, Calcutta, and Penang, Malaysia, by that date as well. These gardens, in conjunction with their counterparts in Europe, formed a critical part of the colonial infrastructure because of their role in moving tropical crops. A curious and lasting effect of this botanical commerce is a pattern of cultivation in which economic production usually occurs where the plants are not native. South American rubber is grown mostly in the Malaysian archipelago, South American cocoa in west Africa, African coffee in the New World tropics, Southeast Asian bananas in Central America, and so on. This pattern had the beneficial—if unintended—side effect of allowing the crops to escape their natural pests and diseases, an effect we still depend on to some degree. (See the discussion of rubber in Chapter 8.)[16]

The botanical garden network was also instrumental in bringing many exotic plants into Europe, and later into North America. No definitive count of the resulting introductions exists, but one survey of central Europe gives a representative picture, at least for the trees and shrubs. Some 2,645 woody plant species were introduced into central Europe from 1500 until the first decades of this century. (See Figure 6–1.) The rate of introductions accelerated steeply from the mid-eighteenth century onward, driven initially by North American and later by Asian imports. Today, the total number of introductions would be considerably higher: some 3,150 exotic woody plants are being grown in German public and botanical gardens alone. At least 210 of these exotic woody plants have become invasive in central Europe. (See Figure 6–2.)[17]

By the time Schieffelin's starlings arrived in New York City, acclimatization was already withering away as a serious interest in European scientific circles. The field had gained much of its intellectual appeal from the great controversy over evolution and religion that culminated in the 1860s. If the form of each creature was ordained, wouldn't its place in creation have been appointed as well? How could a creature thrive in a

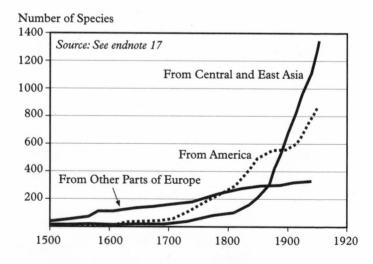

FIGURE 6–1. *Imports of Woody Plant Species into Central Europe, by Place of Origin, 1500–1916*

place where the Creator Himself had not put it? Once the power of evolutionary theory had been established, such questions seemed less and less worth asking. And new realms were opening up. Physiology—the landscapes of the cell—had begun to eclipse the landscapes of forest and field.[18]

By the turn of the century, acclimatization had slid into nearly universal scientific disrepute. Its last gasp as a social force seems to have been in the Soviet Union of the late 1920s and early 1930s, when it formed a part of Stalin's plan for the "Great Transformation of Nature." The plan involved the release of various exotics, including large numbers of North American muskrats and other fur-bearing mammals, once again to the great detriment of the native fauna.[19]

Yet in some senses, acclimatization is still with us,

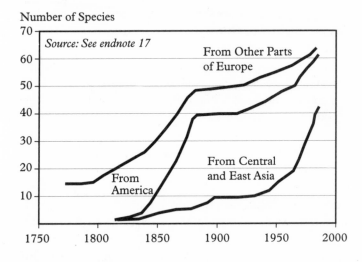

FIGURE 6–2. *Successful Invasions of Woody Plant Species into Central Europe, by Place of Origin, 1787–1990*

not as a social enterprise, but as an assumption that shapes the way we see nature—and as a de facto part of the landscape. Modern sport fish introductions generally make as little ecological sense as the songbird introductions of Schieffelin's era, and have probably done a good deal more damage. (See Chapter 4.) Game introductions, too, have continually repeated the acclimatizers' ecological myopia, although completely novel releases are a rare event now. But the legacy of past releases lives on, especially in the United States. In Texas, for instance, a couple of Eurasian deer species are outcompeting native deer. In the Olympic peninsula of the Pacific Northwest, an introduced herd of mountain goats is damaging delicate subalpine flora. And in the U.S. Midwest, the Asian ring-necked pheasant is competing with the native quail and prairie chicken.[20]

And Stalin was not alone in his fondness for fur. Several mammals have been introduced in various parts of the world to build fur industries: the North American mink in the United Kingdom, the South American nutria (a kind of large rodent) in the U.S. Southeast and Pacific Northwest, the North American muskrat throughout Eastern Europe, the North American beaver in Tierra del Fuego, and the Australian possum in New Zealand. (See Chapter 5.) In all these cases, the mammals have become wetland or forest pests.[21]

But horticulture is the principal means by which acclimatization reaches into the present. To those who do not practice it, gardening may seem merely a coterie obsession, like playing bridge or collecting stamps. But the age and ubiquity of the practice show its profound importance, for both the landscape and the human psyche. When the European colonial powers extracted, in effect, a tribute of plants from their dependencies, they were doing what imperial societies generally seem to have done. There is, for example, a surviving letter from Darius the Great of Persia (who reigned 521–486 B.C.) commending an estate manager for introducing "trees and plants from beyond the Euphrates." In his voluminous and chatty *Natural History,* Pliny the Elder (23–79 A.D.) shows us a Rome as acquisitive of foreign plants as it is of foreign gods. And in the early fourteenth century, Marco Polo describes Kublai Khan's interest in evergreen trees: whenever the Great Khan encountered one he liked, he had it dug up and hauled, by elephants if necessary, to his arboretum in what is now Beijing.[22]

Modern gardeners are still served by the horticultural infrastructure that developed during the Age of Acclimatization. Although the botanical garden network now focuses primarily on issues of critical importance to botany and plant conservation, new garden introductions still emerge from it. The garden network also has an extensive for-profit arm: the nursery industry, which as every avid gardener knows is always looking for new introductions. Plant breeders for the big nursery companies are constantly combing the genome of established garden plants for new varieties, just as plant explorers continue to comb the countryside for entirely new species.

The result is that horticulture—a minor industry in terms of its economic size—is a gargantuan engine of biotic mixing that has helped unleash some of the world's worst plant invasions. (See Table 6–2.) One global survey of 1,060 woody plant invasions, for example, found that in the 624 cases in which the origin of the invasion could be ascertained, 59 percent

Table 6–2. Garden Variety Monsters

Pest	Description
Rubber Vine	Introduced from Madagascar as an ornamental and possible rubber source into northern Australia at the turn of the century. In a reversal of the usual scenario, this island invader of a continent now infests some 350,000 square kilometers of tropical Queensland, where it chokes out native grassland and forest, smothering trees up to 30 meters high.
Clematis Vitalba	This ornamental vine from northern Europe had escaped from gardens in New Zealand by the 1930s. It is doing to that country's forests what rubber vine is doing to northern Australia. When a stand of smothered trees collapses, clematis blankets the resulting tangle, forming mats over a meter thick and preventing any regeneration.
Water Hyacinth	A South American aquatic plant introduced during the nineteenth century into the southern United States, Africa, and southern Asia. Its original use was often as a pool ornament; subsequent uses have included fodder, green manure, biogas production, and wastewater treatment. But given the number of lakes and rivers that have disappeared beneath it, many water managers would be glad to get rid of it without using it at all.

came from botanical gardens, landscaping, or other amenity purposes. In the continental United States and Canada, garden introductions are estimated to account for about half of the 300 or so really serious pest plants of natural areas. Even species like purple loosestrife—species that are proven hazards—generally remain in the trade.[23]

More than 60 percent of North America's worst wild-land weeds are still being sold by nurseries. And new garden invaders continue to surface regularly. In early

Pest	Description
Purple Loosestrife	This European wetland plant probably first reached North America at the end of the eighteenth century in wool imports and solid ship ballast; during the nineteenth century it was imported for ornamental and probably for medicinal purposes. It has now overrun more than 600,000 hectares of North American temperate and boreal wetland, where it has eliminated native vegetation and ruined the waterfowl forage base.
Knotweeds	Bamboo-like plants from east Asia introduced into Europe and North America during the nineteenth century as ornamentals and, in Europe, for game forage. On both continents, knotweeds are outcompeting native riverside vegetation and exacerbating floods by choking off water courses.
Saltcedars or Tamarisks	A group of scrubby, feathery-leaved trees from Central and East Asia introduced into the western United States beginning in the early nineteenth century, primarily as ornamentals but also for erosion control along rivers. Today they infest more than 600,000 hectares along rivers and streams, forming dense thickets of little wildlife value and often eliminating surface water. In the U.S. southwest, it is estimated that saltcedars now absorb a greater quantity of water every year than is used by Los Angeles and all the other cities of southern California combined.

SOURCE: See endnote 23.

1998, for instance, a new weed emerged in New Zealand: a garden escape called holly leaved senecio, a South African daisy that produces purple-pinkish flowers and can grow as tall as a person. Giant purple daisies—one more flower for the chaotic world garden.[24]

7

Accidents

A ship comes into San Francisco Bay in the fall of 1998. It's a container vessel loaded with big metal boxes that cranes will transfer to the flatbeds of tractor trailers or railcars. This ship has arrived from Pusan, South Korea, but some of its containers were loaded at previous ports of call, during a commercial odyssey that took it all the way up the Asian coast from Malaysia. Soon they will be distributed throughout the bay area and far beyond. What's on board? Probably some of just about everything that the West buys from the East: car parts, electronics, furniture, textiles, maybe even chemicals.

What else is on board? No one will ever know—at least as far as this particular ship is concerned. But the immense 112,000-hectare bay is teeming with the results of its ship traffic. Nearly 30 species of exotic fish are now major predators in the bay—a bizarre mixture of East Asian gobies, Atlantic shad, Mississippi catfish, common carp, various species of bass, perch, and sunfish, even goldfish. This crowd of exotic fish has suppressed native fish populations and contributed to the extinction of at least one native species: the thicktail chub.[1]

The European green crab appeared in the bay around 1989. The green crab grows only to about 8 centimeters in width, but it is belligerent, adaptable, and a kind of living vacuum cleaner. It will eat just about anything—shellfish, barnacles, algae, snails— even other crabs. In the 1950s, it invaded and destroyed the softshell clam beds on the coast of New England and Atlantic Canada; subsequent efforts to control it have met with mixed success. Scientists fear a repeat performance on the West Coast.[2]

A couple of years after the green crab arrived, the Chinese mitten crab appeared. (The popular name comes from the fact that its claws are sometimes enveloped in "mittens" of hair.) The mitten crab digs tunnels more than a half-meter deep. Crab colonies can turn shorelines into a honeycomb of tunnels, and scientists worry that the crab will erode the banks of tributary rivers, and even cause levees to fail. In China, the crab is host to an occasionally fatal human parasite, the Oriental lung fluke, which has infected millions of people. The fluke is apparently not present in the bay's crab population, but if travelers ever bring it into the area, it will find a ready reservoir of potential hosts.[3]

An Asian clam (*Potamocorbula amurensis*) arrived in the bay in the mid-1980s and is now the dominant bottom-dwelling organism, occurring in beds at sustained densities above 2,000 per square meter. Like the zebra mussel (see Chapter 4), the clam is an extremely efficient filter feeder; in the shallows of the San Francisco Estuary, it filters the entire water volume at a rate of nearly 13 times a day, which exceeds the reproductive rate of most forms of plankton. The consequent disappearance of the plankton has suppressed shrimp and fish populations—including the local Chinook salmon runs. The clam is also, like the zebra mussel, bioaccumulating dangerous pollutants. Selenium, a toxic substance released into the bay from farm runoff and oil refineries, is being absorbed by the Asian clam at rates two or three times greater than by native clams. Higher selenium levels may be causing reproductive problems in the waterfowl that feed on the clams.[4]

Most of the bay's other creatures—sponges, seasquirts, crayfish, mussels, marine worms, barnacles, sea anemones, the cordgrass in the saltmarshes—are now exotic. Even the plankton may be largely exotic by now, but plankton taxonomy is not well enough understood to say for sure. Thus far, 212 exotics have been identified in the bay. At least another 123 species are of unknown origin, and another exotic establishes itself, on average, every 12 weeks. In some parts of the bay, all the species are now exotic: the bay invites us to think in terms not of exotic species, but of entire exotic communities. San Francisco Bay is ecological chaos: its past is no longer a reliable guide to its future.[5]

$$\star \quad \star \quad \star \quad \star$$

Many of these creatures arrived in the bay by being pumped out of ships' ballast tanks. Despite its apparent stability, a big ship does not sail simply by virtue of its design, any more than an airplane flies simply because it has wings. Keeping a ship upright takes ballast water—lots of it. Moving that water in and out of the cavernous tanks designed to hold it is as critical a part of the nautical routine as managing the rudder or the engines. Ballast water must be taken on when cargo is unloaded, or as fuel is consumed, or to provide extra stability in heavy weather, or sometimes to make the ship ride low enough to pass under a bridge. And for every reason that it is pumped aboard, there is a corresponding reason for pumping it out—taking aboard cargo, making the ship ride high enough to move into a shallow harbor, and so on.[6]

The ballast capacity of a big tanker can exceed 200,000 cubic meters—enough to fill 2,000 Olympic-sized swimming pools—and its pumps can move that water at rates as high as 20,000 cubic meters an hour. That is not gentle suction. Most ballasting is done around harbors, in shallow water, and ships sometimes scour the bottom as they are ballasting up. In the resultant turmoil, the pumps can inject into the tanks a slurry containing hundreds of cubic meters of sediment—along with any small creatures that happen to be in the water or mud, or on nearby harbor pilings.[7]

The tanks of a large ship may come to support a chaotic but more-or-less permanent living community. A large ballast tank can only be emptied completely by opening it up, and that is not done except during drydock overhaul, which on a well-maintained vessel might occur every three to five years. Routine use always leaves plenty of room for biological activity. In one recent survey of large ships reporting no ballast on board, the burden of unpumpable water and sediment in their "empty" tanks averaged 157.7 tons.[8]

Ballast water is a soup stocked from harbors all over

the world. (See Table 7–1.) The holes in ballast intake grates are usually about a centimeter wide. That is probably plenty of room for most marine organisms in larval form—most fish larvae, for instance, are likely small enough to pass through. Sometimes the grates fall off, allowing much larger creatures to enter. In

Table 7–1. Ballast Water Invaders

Source	Description
From Europe	The zebra mussel, a shellfish native to the Caspian Sea region, is rearranging the ecology of many North American waterways.
	The ruffe, a European fish, has established itself in the Great Lakes and is beginning to outcompete the yellow perch and walleye—native fish species with an economic value of $90 million annually.
	A Mediterranean fan worm now forms a kind of living carpet over parts of the floor of Port Phillip Bay, on the southeastern coast of Australia, to the detriment of the local scallop fishery.
From the Americas	Leidy's comb jelly, a jellyfish native to the Atlantic coast of the Americas, has devastated the Black Sea fisheries.
	The American razor clam, a shellfish native to the North American Atlantic coast, has established itself along the western and northern European coasts.
	A bristle worm, also native to the North American Atlantic coast, now constitutes 97 percent of the biomass of the large bottom-dwelling species in the immense Vistula lagoon, on the coast of Poland.
From East Asia	The Chinese mitten crab and the Asian clam have invaded San Francisco Bay.

April 1995, for example, the tanks of a ship that had come into Baltimore harbor from the eastern Mediterranean were found to contain more than 50 healthy mullet 30–36 centimeters long.[9]

But some of the most significant stowaways are microscopic: in 1991, ballast discharge from ships arriving in Peruvian ports from South Asia is thought

Source	Description
	At least eight species of East Asian copepods are established on the Pacific coast of the Americas.
	A starfish *(Asterias amurensis)* from the Northwestern Pacific has invaded the Tasmanian coast; it is threatening local shellfish industries and endangering the native spotted handfish—which could become the first known marine fish to go extinct during the historic era (although there have likely been undocumented extinctions).
	Poisonous dinoflagellate "red tide" plankton, native to the waters off Japan, occasionally shut down oyster farms along the southeast coast of Australia.
	Various seaweeds native to the Japanese coast are established in the waters off Tasmania and along both coasts of the United States.
From South Asia	The Indian bream, a fish, is established in western Australia.
	Another fish, the Indo-Pacific goby, is established in Nigeria, Cameroon, and the Panama Canal.
	A crab *(Charybdis helleri)*, native to the Indo-Pacific and established in the eastern Mediterranean, has now appeared in the waters off Cuba, Venezuela, Colombia, and Florida.
From Australia	An Australian barnacle is outcompeting native barnacles over large stretches of European coast.

SOURCE: See endnote 9.

to have unleashed the first cholera epidemic that the western hemisphere has seen for more than a century. The outbreak may have infected several million people and killed 10,000 of them. Ships outbound from Latin America were found to have cholera-laden ballast upon arrival in ports in Australia and the United States. More cases of imported cholera were observed in the United States in 1992 than in any year since surveillance for that disease began.[10]

There are more than 28,700 vessels in the world's major merchant fleets, and they make up by far the largest part of the world's trading infrastructure. About 80 percent of the world's commodities travel by ship for at least part of the journey to their consumers, and the volume of seaborne trade is climbing steadily upward. From 1970 to 1996, the trade nearly doubled (it climbed from 10,654 billion ton-miles, the standard industry measure, to 20,545 billion ton-miles). Through its ballast systems, the world merchant fleet has in effect superimposed a second set of currents on the world's oceans, and these meta-currents are far more efficient at transporting life over long distances than are the natural ones.[11]

On any given day, the meta-currents are moving perhaps 3,000 different species of every conceivable ecological function: green plant, pathogen, parasite, herbivore, carnivore, scavenger. Nor do the meta-currents reach only harbors. Because of the pervasiveness of ship traffic and the complexities of managing ballast, no coastal site can be considered immune to their effects. But it is in harbors that we can most readily see the implications for the world's coastal regions, which are the most biologically productive parts of the oceans. And in harbor after harbor, the same species are appearing over and over again—the same crabs and clams, the same worms, sometimes even the same fish. The world trading system is creating an extra-geographical marine biota.[12]

Many of the world's worst aquatic invaders have been ballast water releases—the zebra mussel, for example, or that little luminescent blob known as Leidy's comb jelly. Usually smaller than your thumb, this jellyfish is native to the East Coast of the Americas. It was apparently pumped out of a ballast tank and into the Black Sea around 1982. Leidy's comb jelly eats the myriad tiny animals collectively known as zooplankton. And since nothing in the Black Sea will eat it, it launched one of the most intense marine invasions ever recorded. By late 1988, a single cubic meter of Black Sea water could contain as many as 500 of the little jellies. If all the jellyfish could have been hauled out of the sea that fall and weighed, the take would have come to between 900 million and 1 billion tons—at least 10 times the total world fishery catch that year. But the anchovies and other fish that account for the sea's traditional catch had largely disappeared. The jelly apparently provoked the collapse of the Black Sea ecosystem.[13]

In effect, the little predator had snapped what was already a badly rusted chain. During the past several decades, the sea had grown steadily more polluted from fertilizer runoff and the sewage of some 170 million people. This nutrient-rich pollution was feeding clouds of algae, which were robbing the water of light and burning up oxygen as they decayed. The Black Sea is naturally anoxic (suffering severe oxygen deficiency) to begin with. For millennia, rafts of plant material would sweep in from the Danube and the other tributary rivers, consuming the oxygen as they rotted and leaving

only a film of aerated water riding a vast anoxic pool—
the largest such pool on Earth. Now the algae were suffocating that upper layer of life. They were also shading
out the huge shallow-water seagrass beds that had once
functioned as the sea's "lungs"—and as prime habitat
for fish, crustaceans, sponges, and many other creatures.
But zooplankton eat algae—and zooplankton were just
about all that remained of the sea's battered immune
system. Then the jelly ate virtually all the zooplankton.
Algae and jelly were almost the only living things in the
water. At its peak, the jelly alone accounted for 95 percent of the sea's entire wet weight biomass.[14]

By the mid-1990s, the jelly was showing signs of hav-
ing exhausted its larder. Its Black Sea population has
declined, but by 1992 it had invaded the Sea of
Marmara, below the Bosporus, and it has turned up
further south, in the Aegean, as well. Eventually, perhaps, it may infest much of the Mediterranean coastline. Shipping could also take it north, up the great
European rivers that run into the Black Sea, and into
the Baltic. In the meantime, several jellyfish native to
the Black Sea have established themselves in the
Chesapeake Bay, on the U.S. East Coast, and in San
Francisco Bay.[15]

Despite their capacity for havoc, most ballast water
invaders do not get much press. To the news media and
probably to the public in general, marine pollution usually means oil spill. The 1989 Exxon Valdez spill in
Alaska's Prince William Sound, for example, attracted
major media coverage for months. But what about the
spill that is spreading through the sound today? In
December 1997, the U.S. Fish and Wildlife Service
announced that it had discovered four new species of
zooplankton there. They were released into the sound

from tanker ballast and appear to have come from East Asia via San Francisco Bay. Scientists are concerned that the invaders may develop a taste for the same foods that are needed by the Dungeness crab, an important fishery species. As more and more Alaskan oil is pumped, some scientists fear that ballast releases like these could become a general threat to the state's fisheries. The biotic spills, in other words, could become a far greater danger to Alaskan coasts than the oil spills. After all, oil spills may be a grave environmental insult, but they eventually go away. Biotic spills do not.[16]

★   ★   ★   ★

It is apparently a universal dimension of culture, the essential human enterprise, to be riddled with hundreds of these more-or-less accidental "pathways"—thoroughfares, often invisible to their human contemporaries, that have conveyed other organisms far beyond their natural ranges. The industries of the past—some of them still with us, some largely forgotten—have left an enormous biotic legacy in this way.

Some 500 exotic plant species growing around the French city of Montpellier, for instance, have been attributed to wool imports: for centuries, wool scouring was an important local industry. About a century ago, the oyster industry inadvertently released the North American Atlantic cordgrass onto the coast of the Pacific Northwest, in shipments of oyster "seed." In its new range, the cordgrass has converted extensive tracts of tidal mudflats—essential for many bird and fish species—into much less productive marsh. A failed soap-making industry helped the highly invasive Chinese tallow tree into the forests and wetlands of the U.S. Southeast. (The tree's seeds can be made into soap; unfortunately, since the tree is also beautiful, there is another pathway through which it moves: horticulture.) The failed enterprise that has probably left the greatest biotic scar was Leopold Trouvelot's tragic ambition to produce silk in Massachusetts by importing the gypsy moth. (See Chapter 3.)[17]

Many older pathways continue to function, of course, but in terms of aggregate effect, contemporary culture has moved an enormous distance from even the relatively recent past. And it has done so along several dimensions at once: modern industries seem to be opening up more pathways, which are moving a greater volume and variety of organisms at ever-increasing speeds. This unintentional biotic leakage is now probably the fastest-growing category of bioinvasion, at least in the industrial world. In a survey of dangerous exotics thought to have entered the United States from 1980 through 1993, for example, researchers found that among those whose pathway could be identified, 81 percent were accidental introductions.[18]

The forest products industry is a prime example of the trend. The movement of forest products has always entailed serious ecological risk—a shipment of veneer logs, for example, brought the Dutch elm disease to North America. (See Chapter 3.) But those risks have increased enormously as the trade in raw wood products continues to grow, both in volume and in the number of trade routes involved. From 1970 to 1994, the most recent year for which figures are available, export volumes of raw logs increased 21 percent (to 113.4 million cubic meters). Trade in sawnwood nearly doubled (to 108 million cubic meters). Among the countries that have substantially increased their raw log exports in recent years are Chile, China, Ghana, New Zealand,

Papua New Guinea, Russia, and the Solomon Islands. Bigger importers include China, South Korea, Taiwan, and a number of African countries.[19]

What the logs contain besides wood is anybody's guess. At a conference on the dangers of raw wood imports in 1996, a former Oregon Department of Agriculture inspector recalled opening up the hatches of a wood chip carrier that had just arrived from Brazil and watching "a cloud of insects" escape. According to many experts, North America would risk a disaster of epochal proportions if lumber companies in the U.S. Pacific Northwest succeed in weakening federal regulations on importing raw logs. Relaxing the regulations would make it easier to feed the region's overcapacity sawmills, but it would also greatly increase the odds of hundreds of new forest pests eventually making their way into western North America.[20]

Some mill owners, for instance, have been promoting the idea of importing Siberian logs. (Strictly speaking, it is already legal to do this, but no company has found an economical way to handle the required treatment procedures, which involve debarking, heating, then storing the logs in sanitary conditions until shipment.) A U.S. Forest Service inventory of organisms associated with Siberian larch, a major timber species in eastern Russia, turned up 175 species of arthropods (insects and their relatives), nematodes, and fungi, including a Eurasian spruce bark beetle that occasionally causes outbreaks that kill millions of trees in its native range. If weakened regulations were to allow the beetle in, the results, according to the Forest Service, could be "as disastrous for North American spruce as the Dutch elm disease was to elms."[21]

But not all biologically dirty industries are as large as

the forestry sector. Some relatively small industries are playing outsized roles as biopolluters—horticulture, for example. (See Chapter 6.) In addition to escaped garden plants, horticulture releases a large number of weeds, insects, slugs, and pathogens. Historic nursery trade contributions to North America include the chestnut blight, the white pine blister rust, the balsam woolly adelgid (a serious pest of the eastern hemlock), and the beech bark disease complex.[22]

The pet trade, the animal equivalent to horticulture, is another major conduit for biotic spills. The domestic cat, for example, is a formidable and now nearly universal predator: wild cats appear to be a serious stress on bird populations in Europe and North America and on small mammal populations in Australia. And there are plenty of other pet predators out there as well. The reptile trade has brought Florida some 20 species of exotic lizards, some of which prey on native lizards. In Hawaii, escaped chameleons are competing with native birds for insects.[23]

By far the most ecologically disruptive sector of the pet industry is the aquarium trade. In increasing numbers, aquatic plants, snails, shrimp, fish, and various other denizens of hobby aquariums are finding their way into natural waters. Some escape from breeding facilities; others are offered their freedom by soft-hearted but misguided owners who have tired of their charges. The consequences in some cases have been dreadful. Hydrilla, a popular aquarium plant from South Asia, escaped from a culture facility in Florida in the early 1950s and is now a premier aquatic weed throughout the Southeast, as well as on much of the West Coast. Hydrilla is clogging more than 40 percent of Florida's rivers and lakes.[24]

Various species of aquarium fish—and the collection includes a substantial share of the world's tropical freshwater fish—are rapidly approaching cosmopolitan status (occurring virtually everywhere there is available habitat). Such standard ornamentals as guppies and swordtails, both native to central America, can now be found in tropical ponds and streams all over the world, especially near cities. Some highly disturbed but relatively clean habitats, like the canal systems of central Florida or the streams on the Hawaiian island of Oahu, have in effect become giant natural aquaria; aquarium species now dominate.[25]

Healthy aquarium fish command a good price, and efficient transport makes it possible to sell to distant markets, so breeders have set up shop throughout the warm regions of the world. If it is beautiful and breeds easily, it has a good chance of making a new life in India, Malaysia, Thailand, the southeastern United States, and various places in between. Of those exotic fish species established in the United States that are completely foreign to the country, about 65 percent arrived through the aquarium trade. And new ones are establishing themselves all the time. At the time of writing, the latest addition is an Asian eel that has taken up residence in ponds near the Chattahoochee River, a major southeastern river. Like the Asian walking catfish that has invaded Florida, the eel is an efficient predator that can breathe both air and water, and it can move overland from one pool to another.[26]

The aquarium trade is spreading fish pathogens as well. Sanitary screening has turned up at least 42 diseases in aquarium fish intended for shipment to Australia. In that country and elsewhere, some of these pathogens have escaped into the wild; a few apparently owe a nearly worldwide distribution to this industry.[27]

Nor is it just amateurs and their suppliers who are responsible for these releases; scientific aquariums occasionally commit such blunders as well. The most notorious research escape is a Pacific seaweed, *Caulerpa taxifolia*, that apparently was accidentally released into the Mediterranean from the Oceanographic Museum in Monaco about 15 years ago. Since then, it has turned some 3,000 hectares of the Mediterranean and Adriatic seafloor into what James Carlton, an expert on marine invasions, describes as an aquatic version of Astroturf.[28]

⋆   ⋆   ⋆   ⋆

A great deal of biopollution cannot really be attributed to any particular industry. All sorts of pathways, for example, have unfolded through the general infrastructure of movement. Roads and railroads carved pathways for exotic weeds into forest and prairie. Canals have been highly efficient corridors of invasion when they connected bodies of water that had previously been distinct.

The Erie Canal, for example, breached the Allegheny Divide in the U.S. Northeast, and allowed more than a score of fish and mollusks native to the Mississippi–Great Lakes basin to establish themselves in the Hudson River drainage. The Welland Canal around Niagara Falls probably admitted the sea lamprey into the upper Great Lakes. (See Chapter 4.) And the Suez Canal, a conduit for 20 percent of world maritime traffic, rejoined the Red Sea to the Mediterranean in 1869, after some 20 million years of separation. Thus far, nearly 300 exotics are thought to have found their way into the Mediterranean through the Suez, including the Red Sea jellyfish, which now produces mass summertime swarms along the sea's eastern coast.[29]

Ballast releases like the Leidy's comb jelly have many precursors as well, although their effects now often seem perfectly "natural." During the last century, wooden-hulled ships apparently brought a tiny, wood-boring crustacean (the isopod *Sphaeroma terebrans*) to the Atlantic coasts of the Americas from the Pacific. *S. terebrans* has spread throughout Atlantic mangrove communities—the fringe of stilt-rooted trees common along tropical and warm temperate shores and essential for maintaining the wildlife value of those ecosystems. The isopod attacks and kills mangrove root tips, thereby controlling the trees' seaward spread. In the words of James Carlton, who discovered the invasion, *S. terebrans* has "virtually 'reset' the seaward history" of the western Atlantic's mangrove ecosystems. But apparently it took a century or so for anyone to notice.[30]

In Europe, a fungal pathogen of North American crayfish was accidentally released into Italian waters around 1860 and is inexorably erasing the native crayfish from that continent. Probably very few Europeans know that most of the crayfish in their streams are exotic North American species, introduced because they are resistant to the fungus.[31]

But as with particular industries, the invasion potential of the trading network has been expanding radically in several dimensions at once. In terms of volume, speed, trade routes, and the variety of organisms involved, the modern trading system dwarfs anything previous eras have seen. Take volume first: at the turn of the century, a substantial ship might have a capacity of 3,000 tons; by World War II, 10,000-ton ships were common; today, ships are often 150,000–250,000 tons, and the largest vessels exceed 600,000 tons. The sheer increase in size may help explain why ballast water invasions seem to have gone from a dribble to a torrent somewhere in the 1970s or early 1980s.[32]

For many types of organisms, greater speed may be more important than greater volume, since the shorter the time in transit, the higher the odds of survival. For ships, the quantum leap in speed actually occurred in the latter half of the last century. In 1851, the William T. Coleman California Line was billing its *Syren*— "The A 1 Extreme Clipper Ship"—as capable of sailing from San Francisco to Boston (via Cape Horn) in 100 days, and from Calcutta to Boston in 96 days. It would have taken the *Syren* four to six weeks to cross the Atlantic. The transition from sail to steam cut that time in half. By the time Eugene Schieffelin's starlings shipped out in 1890, steamers were crossing the Atlantic in only two weeks—considerably better odds for even a hardy land bird like the starling. Land transportation underwent a revolution at about the same time. By the 1870s, North America had been girded by rail and it was possible to travel from one coast to the other in a couple of days. That kind of speed, along with the ubiquitous water towers that were an essential part of transportation infrastructure in the steam age, gave a tremendous boost to fish introductions.[33]

Air traffic, of course, represents another quantum leap in speed, and air cargo is a rapidly expanding sector in the trade network. Air cargo traffic is growing at about 7 percent a year (in terms of ton-kilometers). In 1989, only 3 airports received more than a million tons of cargo; by 1996, that number had risen to 13. Many organisms that would die or be detected during a shipboard crossing can travel easily by air, including human pathogens (as discussed later). Anyone infected with cholera who boarded the *Syren* in Calcutta would be very sick long before reaching Boston. But cholera and many other diseases are bound to be undetected frequent fliers on dozens of air routes.[34]

The types of pathways within the trading system are changing as well. Older pathways have narrowed or disappeared, while new ones have sprung into existence. Some of the most important changes involve ship design. In the days of sail, for example, there were no ballast tanks. Ships were ballasted with tons of dirt, stones, and chunks of iron laboriously shoveled in and out of their holds. Portside "ballast heaps" became home to all sorts of exotic plants, earthworms, beetles, and other small denizens of the soil. And the wooden hulls themselves had enormous biological potential. Seaweeds and barnacles anchored themselves to the hulls; fish, shrimp, crabs, and other little animals took up residence among them. The resultant fouling communities, teeming with all sorts of life, could be more than a meter thick. Specialized boring mollusks called shipworms chewed directly into the hulls. Their continual gnawing opened up little caverns that became home to other travelers.[35]

Today, solid ballast is a thing of the past, and hull-fouling communities are much sparser, thanks to toxic antifouling paints, high speeds, and port times often measured in hours instead of months. But inside a ship, there is still plenty of room for stowaways in bilges, sea chests (the ports leading to the ballast tanks), nets, chain lockers—and in some places an ancient mariner would never dream of. In the U.S. Pacific Northwest, a series of portside infestations of the Asian strain of the gypsy moth were detected in 1991 and apparently eliminated during the following year. (Thus far, only the European strain of the moth has established itself in North America; the Asian strain is more mobile and therefore much more dangerous.) The pathway, it turned out, was the lighting aboard grain carriers coming in from East Asia. The ships' lamps produced light of a wavelength that strongly attracted the moths, and the fast crossing times—so important in reducing hull fouling—helped insure the moths' safe arrival.[36]

Another major leap in invasion potential involves containers, those big metal boxes that have revolutionized the freight industry during the past couple of decades. Containers are nearly ubiquitous: they move by ship, rail, and road. They can be stacked for weeks or even months in ports or railyards, allowing plenty of time for pests to enter. They offer a safe haven to anything that manages to get inside, since they are very difficult to inspect. They are rarely cleaned between shipments, and they may not be unpacked until they are hundreds of miles from their ports of entry. Containers have integrated sea and land transit. They have broken the old link between shipborne exotic and portside invasion.[37]

The overwhelming share of world shipping volume is in bulk commodities, like grain or oil. Materials of this sort cannot be containerized, but almost everything else can be. And increasingly it is. In 1995, the most recent year for which figures were available, world container traffic had reached 135 million twenty-foot equivalent units (because containers come in several sizes, total volume is measured by an abstract unit rather than by the absolute number of containers). As a share of total shipping volume, container traffic is growing steadily; it rose from a mere 1.6 percent in 1980 to 6.4 percent by the end of 1996—a fourfold increase. And the container is becoming a major means of linking developing-world economies with their industrialized trading partners: by 1996, slightly more than half of world container volume was passing through the ports of developing countries.[38]

Containers have been identified as significant pathways for insects, weed seeds, slugs, and snails. Probably the most dangerous exotic known to be using this pathway is the Asian tiger mosquito. For decades, this mosquito had been a common pest throughout much of the Indo-Pacific region, from Madagascar all the way to Hawaii. Then in the mid-1980s, it embarked on its version of world conquest, largely by riding in container-loads of used tires. A tire with a little water in it is ideal mosquito habitat, and millions of used tires are traded internationally every year—some to make recycled rubber products or asphalt, but most as fuel for power plants, cement kilns, and so forth.[39]

The Asian tiger mosquito is currently known to have established itself throughout the southeastern United States, Brazil, southern Europe, South Africa, Nigeria, New Zealand, and Australia. It is an extremely aggressive biting pest. In Asia, it is a major vector of dengue fever, an excruciating disease that gets its common name—break-bone fever—from the pain it inflicts. Dengue infects about 560,000 people each year and kills 23,000. The mosquito can carry at least 17 other viral diseases, including various forms of encephalitis and yellow fever. It may have been a factor in the 1986 yellow fever epidemic in Rio de Janeiro, in which about 1 million people were infected. In 1991, it was discovered in the midst of a yellow fever outbreak in Nigeria. Around the same time, it was a suspected vector in several encephalitis epidemics in Florida.[40]

Researchers fear that the continued spread of the mosquito could substantially increase the disease burden of both the industrial and the developing world. Even where the diseases it carries are already well established, its arrival could be trouble. In the Caribbean, for example, dengue fever is primarily a disease of the cities because the virus still lacks an efficient rural vector—a role this mosquito could readily play.[41]

Because the world trading system is so large and complex, the pathways within it are in a continual state of flux. Like vessels in some sort of global capillary network, they are constantly growing and constantly dissolving, only to rebuild themselves elsewhere. As developing countries, for example, trade increasingly among themselves, it is likely that whole new sets of pathways will open up—between India and Africa, for example, or Latin America and Southeast Asia. The geographical shifting, along with the complexity of the pathways themselves, gives the "ecology" of the world trading system a kind of demented complexity, like a Rube Goldberg drawing. Who would have thought, prior to the early 1980s, that rubber recycling would be instrumental in spreading a mosquito? Or that manufacturers of a certain kind of light were unwittingly increasing the pest risks to North American coniferous forest?[42]

In terms of immediate social effect, the most important set of pathways involves the growing movement of humanity itself, which is increasingly stirring the world's human pathogens into a single, integrated, microbial system. No previous era has experienced such an uproar of human movement. Every week, about 1 million people move between the industrial and the developing worlds; *every day*, about 2 million people cross an international border.[43]

Travel and tourism is now probably the world's biggest industry in terms of its annual receipts, which amount to more than $3.4 trillion. World air passenger traffic—the best single indicator of long-distance travel—is increasing at about 6 percent per year. By 2000, the civilian world air fleet will be moving more than 1.7 billion passengers annually, 522 million of them on international flights. People infected with serious communicable diseases are presumably moving through this system all the time. Doubtless, many of them have no idea they are infected. In the United States, for example, there are about 1,000 new cases of malaria every year, and nearly all the victims apparently pick up the disease while traveling.[44]

Permanent migration, like tourism, has reached unprecedented levels. Every year, some 110 million people immigrate to another country. In addition to these "standard" immigrants, the stream of international refugees and internally displaced persons has increased almost every year since the end of World War II. Since the beginning of this decade alone, their number has grown by more than 75 percent, from about 30 to 53 million. Many of these people end up in camps or shantytowns that are among the world's most miserable and disease-ridden places.[45]

In general, the world we are traveling through seems to be getting sicker. During the past two decades, some 30 new diseases have emerged—diseases like AIDS, Ebola, and the "flesh-eating" streptococcus. At the same time, several of humanity's oldest and deadliest scourges—malaria, cholera, and tuberculosis, for example—may be gathering strength.[46]

This resurgence of infectious disease is driven by a complex of environmental and social forces.

Waterborne diseases like cholera lurk in the open, untreated sewers used by some 1.7 billion people, mostly in the rapidly growing slums of Third World cities. Other pathogens are spreading because their vectors are on the move—creatures like the Asian tiger mosquito. And a growing number of pathogens—all three of the ancient diseases just mentioned, for example—have evolved drug-resistant strains. (Another reason for the resurgence, however, may be the fact that some 2.5 billion people, about 40 percent of the global population, do not have access to essential drugs at all.) Infectious diseases kill about 16.4 million people every year—about a third of all human mortality.[47]

Human movement is the common denominator within much of this complex: it makes every local misery a global concern. Take the mosquito-vectored disease yellow fever, for example. Yellow fever has two strongholds: the forests of Latin America and the west African countryside. In 1992, for the first time in a quarter-century, the African reservoir of the disease reached east, into Kenya. Many experts fear that the Kenyan epidemic is the beginning of a new conquest. Kenya is a favorite destination for Indian emigrants; there is, consequently, a considerable amount of air traffic between the two countries. Yellow fever is not yet present in Asia, and the Indian population is wholly unvaccinated against it. Some experts regard an Indian epidemic as all but inevitable—especially since another favorite destination of Indian emigrants is Latin America.[48]

Travel not only spreads diseases, it can intensify them. It used to be, for example, that on any particular Caribbean island there was only one type of dengue fever, but travel and trade are mixing the forms of the disease. Infection with multiple strains of dengue produces dengue hemorrhagic fever (DHF), a condition that is far more likely to be fatal than ordinary dengue. Globally, the incidence of DHF has increased nearly 10-fold, to 260,000 cases per year, since 1986. Such overlapping infections may open up whole new dimensions of public ill health. One reason AIDS kills much faster in Africa than it does elsewhere may be that so many of its African victims are also infected with malaria. Malaria (another mosquito-borne disease) already infects 225 million people a year and kills 2 million of them. Perhaps one side effect of its resurgence will be an increase in the death toll from AIDS.[49]

* * * *

Medieval Arabian scholars were able to map the path of the Black Death: from its origin in the central Asian steppes, it followed the trade routes overland to the Crimea, then sailed to Byzantium and the great cities of the eastern Mediterranean. Those chroniclers were watching a particularly horrible form of biotic mixing on a continental level. At the end of the fifteenth century, the Age of Discovery opened, and the biotic turmoil began to unfold on a global level, as the European biota spilled out over much of the Earth. Gradually, during the following centuries, various creatures from other regions were pulled into the flux—South American potatoes, Australian eucalypts, North American salmonids. Today, we have inaugurated a new era of ecological chaos. There is no longer any single predominating current, nor is there any type of organism that we can say with assurance is exempt from movement. Just about anything could be transported anywhere. Who will be able to map the plagues of the next millennium?

8

Economic Invasions

During the summer, some of the wetlands in the U.S. Northeast are cloaked in brilliant purple bloom. It is a spectacular sight, and could easily be mistaken for a perfectly natural one. But this ornament of the marshes is purple loosestrife, an aggressive Eurasian plant that can overrun a North American wetland and reduce its wildlife value to roughly that of a parking lot. If you know what the plant is, there is a good chance you will notice the marsh is missing its waterfowl; if you do not know, you may just enjoy the view. It is possible to look right in the face of a monstrous change and not see it— if the monster does not fit into your way of seeing.

In 1993, the Office of Technology Assessment (OTA), a research branch of the U.S. Congress, issued a report on exotics in the United States. The report contained a kind of baseline estimate of the economic damage done by 79 major invasions into the country during the course of this century: $97 billion. And since that figure only includes losses that OTA was able to document, it is probably just a small fraction of the losses actually inflicted. Yet despite their magnitude, the costs of invasion remain largely invisible. The damage is real enough, but most of it is concealed by the loose-strife factor: invasion is not a conceptual category that figures in economic analysis. Even among economists concerned specifically with environmental problems, exotics rarely get any attention.[1]

Because invasion has yet to be comprehensively studied as an economic process, there is little in the way of hard data to fill this hole in our vision. But we know enough about the general workings of invasion to identify the parts of our social machinery that are most vulnerable to exotics. And one of the weakest spots lies in the economy's "monoculture syndrome."

In any kind of ecosystem, natural or artificial, the closer you get to a monoculture—a system dominated by a single species or variety—the less stable the system is likely to be. Any pest that succeeds in attacking the dominant organism stands a good chance of overrunning the entire terrain. That, of course, is why conventional industrial agriculture is so vulnerable to pests: monoculture is almost always the objective in large-scale farming. And since so many exotic pests are abroad in the agricultural landscape, invasion obviously accounts for a good deal of agriculture's pest losses.

But figuring out exactly how much is usually a matter of loose conjecture. OTA, for example, produced a rough estimate of damage that exotic weeds are doing to

U.S. agriculture. By a very conservative reckoning, U.S. crop losses attributable to weeds run to $4.1 billion annually. Since 50–75 percent of the country's major crop weeds are exotic, their share of the bill would appear to be roughly $2–3 billion. Add in a corresponding percentage of total farm herbicide costs, and agricultural losses to exotic weeds in the United States would seem to lie somewhere between $3.6 billion and $5.4 billion annually. (Incidentally, losses to weeds, frequently the largest category of agricultural losses, are not included in that OTA $97-billion estimate.)[2]

What if this approach were carried to its logical extreme? The total value of the world's eight most economically important crops is roughly $580 billion. Pests may claim as much as 42 percent of the yield, or $244 billion. Factor in annual global pesticide expenditures ($31 billion), and the bill climbs to $275 billion. The percentage of all types of pests that is exotic varies widely from one agroecosystem to the next: it might range from 20 percent to 90 percent. Losses to exotics, in this analysis, would therefore lie between $55 billion and $247.5 billion.[3]

There are some important caveats to this kind of assessment. First, even if exotic pests were to disappear entirely, a substantial portion of the losses currently attributed to them would remain, because native pest populations would expand in the absence of exotic competition. Second, there is no certain correspondence between the percentage of pest species that is exotic and the percentage of damage actually done by exotics because a large share of losses is often the work of only a few species. When the main culprits are exotic, then the true share of exotic damage may be far higher than a percentage estimate would suggest; when they are not, it may be much lower. Third, this approach does not capture "collateral damage"—ecological degradation that is of no immediate relevance to agricultural production.

Even so, it is clear that the damage exotics inflict on the monocultural landscape is enormous, and the system's vulnerability is growing in several dimensions at once. There is, first of all, the basic problem of new pests coming into the system and old ones moving into new areas. If the Mediterranean fruit fly, for example, were to establish itself permanently in the continental United States, it could cost U.S. agriculture $1.5 billion annually. Then there is the pesticide resistance problem, which is also now imposing formidable losses. A resistant strain of bollworm, for instance, forced China's cotton yield down by a third from 1991 to 1993, a blow from which the country's cotton sector has yet to recover fully.[4]

And eventually the costs of genetic erosion (through the loss of traditional crop varieties, or landraces) may also begin to bite. In the late 1980s, the Russian wheat aphid devastated U.S. wheat and barley fields. The United States owes its current wheat production to a breeding program that used genes from Middle Eastern wheat strains resistant to the aphid because they are evolving with it. Sooner or later, the aphid is likely to overcome the resistance locked into the current U.S. varieties. If those landraces have disappeared by that time, then the U.S. wheat crop could be in serious trouble. (These trends are covered in more detail in Chapter 2.)[5]

The best way to reduce pest pressure, according to most advocates of sustainable agriculture, is to build greater diversity into agro-ecosystems. More crops growing in a field—or even more kinds of a single crop—mean that no particular pest is likely to be able to eat the whole field. The result should be a more resilient system, and one that requires less in the way of pesticides.

But the big money in commercial agriculture does not, by and large, appear to be weighing in on the side of diversity. Instead, agribusiness seems to be looking primarily for ways to extend its conventional reliance on pesticides, as for instance by engineering herbicide-tolerant crops. Take the agribusiness giant Monsanto, which leads the field in the development of transgenic crops. Monsanto has engineered varieties of rape (an oilseed crop), corn, cotton, and soybean for tolerance to its top-selling weed killer, Roundup (glyphosate). By 1997, Monsanto had nearly 7.9 million hectares in North America, Argentina, and Australia planted in transgenic crops designed for such traits as herbicide tolerance or production of the *Bacillus thuringiensis* toxin (a natural insecticide; see Chapter 2). Projected plantings for 1998 were in the range of 20 million hectares.[6]

There is nothing inherently wrong with genetically engineered crops—often biotech simply functions as a way of speeding up what conventional breeders do. Some scientists even make an environmental case for herbicide-tolerant crops: such varieties may, for example, permit greater use of "no till" agriculture, a form of farming without plowing that greatly reduces topsoil erosion, another critical agricultural problem. (Plowing, the usual way of breaking up weeds, can cause a great deal of erosion, so the idea is to substitute herbicide for the plow.) But whatever their other virtues, herbicide-tolerant crops are unlikely to supply a remedy for agriculture's massive chemical dependency.[7]

The monoculture syndrome is not limited to agriculture. Exotic tree plantations are also attracting more and more pests. The extreme case is Uruguay, which had to abandon its exotic Monterey pine plantations entirely because of an exotic pest complex. (See Chapter 3.)[8]

Aquaculture, too, has some degree of monoculture vulnerability. The worst case is the $8-billion shrimp farming industry, which is active in more than 50 countries, mainly in southern Asia and Latin America. During the past decade or so, a host of extremely virulent pathogens have been circulating through shrimp operations all over the world. Taiwan used to be the world leader in shrimp production, until it lost 80 percent of its harvest to a viral outbreak in 1987. By 1993 China was the world's biggest shrimp producer, but the country lost virtually all its production during the course of a few days in June of that year to an epidemic complex of viruses, bacteria, and protozoa. At about the same time, an epidemic cost Ecuador's shrimp industry $200 million. India's industry crashed in 1994, when an unidentified disease wiped out $63.8 million worth of shrimp in Andhra Pradesh and Tamil Nadu. A year later, outbreaks were wiping out ponds in Central America and Texas.[9]

But the costs to the industries that work most closely with nature are not confined to monoculture contexts. Even in more-or-less natural environments, such industries sometimes suffer major losses to exotic pests. The Leidy's comb jelly, for example, is thought to have cost Black Sea fisheries $30 million a year directly (see Chapter 7) and is the critical factor in a crisis that has deprived some 2 million people of their livelihood, either in the fishing fleets or in the businesses that served them. Cumulative direct losses to the invasion are now probably in excess of $350 million.[10]

A terrestrial analogue to the overfished and polluted Black Sea might be the overstressed forests of the United States, where exotic pest damage may amount to $4 billion annually. The chestnut blight may have been the region's most costly forest pest, but too many years have passed since the beginning of the epidemic to quantify its effects with any precision. In terms of current measurable effect, the gypsy moth is a better example of the problem. U.S. Forest Service expenditures for controlling this pest currently average around $7 million annually. The most severe outbreak of the moth, in 1981, is estimated to have done $764 million in damage.[11]

★   ★   ★   ★

Invasions impose some of their biggest costs not in the natural world but right in the heart of the economic machine—the pipes, foundations, cables, and other pieces of the "built environment" that collectively make the social world a place distinct from the natural world. Artificial structure is where those two worlds join; it is how society rests itself on the soil and the water, amidst the weather and the ebb and flow of other living things. But that joint is sound only when things on either side remain within their established rhythms. When an exotic changes the ecological ground rules, cracks may begin to form.

The power lines on the island of Guam, for example, were not designed with the presence of the fast-breeding, arboreal brown tree snake in mind. (See Chapter 5.) The snakes find the power poles and wires irresistible; they are continually climbing them, shorting out the system, and frying themselves in the process. An outage can cost the Guam Power Authority thousands of dollars in burned-out equipment and repairs, and the utility's losses are presumably dwarfed by private-sector costs, not just in equipment failures but in ruined refrigerated goods, lost productivity, and so on. Similarly, the housing developments of Hawaii did not anticipate the appetite of the Formosan termite, which causes nearly $150 million in damage annually to private homeowners—or six times total government funding for pest control on the islands. (This termite is now spreading through the southern United States as well; in California, it may eventually cause losses equivalent to 1 percent of the total value of wooden structures throughout the state.)[12]

One of the worst exotic "fractures" is extending through the lakes and rivers of North America, where intake pipes, navigation locks, and other pieces of aquatic infrastructure are being fouled by exotic shellfish, especially the zebra mussel. (See Chapter 4.) Originally from the Caspian Sea region, the zebra mussel spread throughout much of Europe before the Industrial Revolution. A good deal of European infrastructure therefore grew up with it. But North American waterworks did not; they lack the deeper intakes, sand filters, and other features that help reduce this type of fouling.[13]

That vulnerability, combined with the mussel's explosive invasion, has created an acute hardening of industrial arteries, especially in the Great Lakes region. The mussels can plug small intakes completely and reduce the effective diameter of giant ones by more than two thirds. Downtime, retrofitting, poisoning the mussels with chlorine, scraping them out of the pipes, or blasting them out with high-pressure hoses—the costs of mussel control are beginning to add up. In the Great Lakes region alone, documented cumulative losses to major users of untreated water had passed $120 million by the end of 1994. Power plants have been hardest hit, because they need so much water for cooling—in some cases more than 2 billion gallons per day. The most vulnerable of all are nuclear plants, where cooling is a safety issue of the highest order. At least 12 North American nuclear plants are infested by mussels; the cost of mussel control at a nuclear plant averages $825,000 a year.[14]

Retrofitting will eventually reduce these expenditures but it will never eliminate them entirely. And these are only the costs to major users; the costs to all users, from individual boat owners to municipal water authorities, would be far higher. Estimates vary widely, but a representative range for cumulative costs in the near term—say, by the year 2000—would be $3.1 billion to $5 billion.[15]

Exotic plants are plugging up infrastructure as well. Water hyacinth has overgrown dams in Zimbabwe, sometimes backing up enough water to burst them. Occasionally, it threatens Uganda's main generating station at Owens Falls. (See Chapter 4.) And what water hyacinth is doing on the surface of the water, hydrilla is doing below. This escaped aquarium plant is clogging generating stations, municipal water supplies, and harbors in the U.S. Southeast. In 1994, Florida's heavily infested Kissimmee River flooded when mats of hydrilla blocked the drainages into adjoining lakes.[16]

The U.S. West Coast has a serious hydrilla problem too. The same year it caused the Kissimmee to flood, hydrilla was found in California's Clear Lake, part of the Sacramento River drainage and a major link in the state's irrigation system. The annual hydrilla-fighting budget went from about $600,000 to $1.9 million, since this is not a battle the state can afford to lose. California's hydrilla control effort is now essential to maintaining the state's $22-billion agricultural sector. Aggressive, widespread aquatic weeds like these are undoubtedly adding to the cost of managing water in many parts of the world. No definitive estimates are available, but losses in the United States alone are thought to run into billions of dollars annually.[17]

Plant invasions raise the cost of water in another way as well: by pumping large quantities of it into the air. A thick infestation of water hyacinth, for example, can drop the level of a lake substantially as the plants, in effect, drink in the lake and breathe it out through their leaves. A hyacinth infestation has so greatly reduced the area of a major lake in southern China—Lake Dianchi, near the city of Kunming—that the local climate has grown noticeably more arid. (The infestation has apparently also eliminated 38 of the lake's 68 fish species.) In regions that are already arid, the problem can be far worse. Exotic pine and acacia invading South Africa's Western Cape Province are threatening the water supply of Cape Town. One recent study found that if left unmanaged, the invasion could cut Cape Town's water supply by about a third during the course of a century.[18]

Something like that may already have happened in the U.S. Southwest, where dense thickets of Asian saltcedars line rivers and reservoirs. Saltcedars are prodigiously thirsty. One good-sized tree can take in 200 gallons of water a day—enough to run a modest

U.S. household. Saltcedar covers more than 400,000 hectares of the region and may now absorb a greater quantity of water than is used by all the cities of southern California combined, including Los Angeles.[19]

The most spectacular destruction that exotics visit upon the built environment involves fire-adapted vegetation. (See the discussion of cheat grass in Chapter 2.) In areas with rapid natural fire cycles, housing developments are always a serious social gamble, but housing developments planted with incendiary vegetation are the social equivalent of smoking next to the gas pump.

Eucalypts are favorite ornamentals in California, and it is no secret that eucalypts burn vigorously. The 1923 fire in the city of Berkeley, across the bay from San Francisco, burned nearly half the city and was fed in part by eucalypts. So was the 1970 fire, after which city authorities recommended cutting the eucalypts, a suggestion that was largely ignored. In 1992, after five years of drought, the city burned again. The fire consumed nearly 690 hectares, 3,500 houses, and killed 25 people. Total damage ran to $5 billion. According to one study, eucalypts contributed 70 percent of the energy released from the vegetation that burned. In the wake of the fire, authorities developed new regulations on the planting of eucalypts and other fire-adapted plants. But once again, the regulations are being largely ignored.[20]

South Florida has the makings of a similar crisis, involving another Australian tree, the melaleuca. (See Chapter 1.) The melaleuca was introduced into the region around the turn of the century as an ornamental, and in the hope of turning the Everglades into timber-producing forest. Now that the Everglades are a celebrated natural area, the melaleuca has come to be regarded as the tree from hell. It doesn't so much burn as explode—its leaves are laced with flammable chemicals and its layers of papery outer bark are tinder dry. After a fire, which does not usually kill it, the tree rains millions of seeds onto the burnt-over land. A seedling can reach 2 meters in its first year.[21]

Florida's worst melaleuca fire occurred in 1985, around the city of Miami. The 58th Street fire burned for 35 days, consumed 3,530 hectares, and blanketed the state's southeast coast with the thick, oily smoke that is another of the melaleuca's trademarks. The smoke coated electrical lines, causing them to arc; transformers exploded. There were hundreds of traffic accidents. In South Florida's heavily populated Dade and Broward counties, suburbia is rapidly invading areas already thoroughly invaded by melaleuca. The collision between suburban homeowner and pyrotechnic tree is likely to be continual and expensive.[22]

*   *   *   *

Perhaps the biggest social cost that exotics can exact is in the spread of disease. The continual movement of pathogens and their vectors may create a cumulative systemic stress analogous to the invasion pressure on agriculture. Latin America probably provides the best current illustration of the risk. In 1991, epidemic cholera returned to the Americas, likely via ships' ballast. (See Chapter 7.) Peru, which suffered the brunt of the epidemic, lost more than $1 billion in exports of seafood potentially contaminated with the disease and in tourist revenues. During the next four years Latin American governments spent more than $200 billion in emergency repairs to sewage and drinking water systems as they struggled to stem the epidemic.[23]

Cholera has been joined by another old scourge: the yellow fever mosquito, originally from Africa but a well-established pest throughout the tropical and warm-temperate regions. In the Americas, the mosquito has been the target of eradication campaigns for most of this century. That effort achieved its greatest success under the Pan American Health Organization in the early 1960s, when authorities thought they had cleared the mosquito from most of the hemisphere. But program funding dried up in the following decade, and the mosquito is now back with a vengeance. It has returned to all 21 countries from which it was ostensibly banished. Most of the hemisphere's tropical cities are now reinfested.[24]

In the Americas, this mosquito is the principal vector for dengue fever, and since the mid-1980s rates of dengue infection in Latin America have increased 10-fold. As its common name suggests, it is also the main vector for yellow fever (in urban settings), and its return makes the possibility of a major yellow fever epidemic higher than it has been in more than 50 years. Adding to the danger is the arrival of the Asian tiger mosquito, which can also carry both diseases. The Asian tiger mosquito is not likely to be as easily suppressed as its African counterpart was, because it is much more adaptable in its choice of breeding sites.[25]

In addition to the appalling medical crisis, a resurgence of cholera, dengue, and yellow fever would impose a crippling financial burden on the region. Overlapping epidemics would create substantial increases in the demand for emergency medical care, water infrastructure repair, and mosquito control. (Full-bore mosquito control can be expensive: a single county in South Florida, Lee County, spends $9 million a year on it.) At the same time, the epidemics would likely force a substantial decline in two major sources of foreign exchange: food exports (both fishery and agricultural) and tourism.[26]

★ ★ ★ ★

It is not generally useful to blame the patient for the disease, but the global economy is not a passive victim of the exotics that infest it. Mosquitoes and weeds are not created by the economy, but the economy is behind most of the damage that they are doing. And certain kinds of economic behavior continue to set the system up for additional invasion pressure.

There is, for example, a tendency to allow managed invasions to function as a means of redistributing wealth—from an asset held by the general public to the coffers of a particular industry that may profit from the exotic, at least in the short term. Even in agriculture, where the widespread use of exotic crops is essential, this kind of imbalance occurs, although it has not been well studied.

One of the few attempts to quantify the problem is a 1994 survey of Australian forage plant introductions. Of the 466 species included in the study, only 21 turned out to have some actual merit for the livestock industry; 60 were definitely invading rangeland—including 17 of the 21 potentially useful plants. Only 4 of the introductions were beneficial and noninvasive. The ranching industry garners some marginal profit from these introductions (at the rate of about $2 per hectare), while the public picks up the substantial control expenses (at $30–120 per hectare).[27]

Who collects the profits and who bears the costs? Many industries that deploy exotics invite that structural question. Take Lake Victoria, where a local fishing economy—overexploitative though it was—has been replaced by the Nile perch fishery, which is designed primarily for export rather than for meeting local needs. (See Chapter 4.) Tropical pulp plantations have set off a similar dynamic in Latin America and Southeast Asia, where traditional forest economies have dissolved in the face of capital-intensive, export-oriented production. (See Chapter 3.)

With shrimp farms, both the social and natural corrosion is even more intense. Shrimp operations typically destroy coastal mangrove stands, thereby dooming the coastal fisheries, since mangrove roots are fish nurseries. In Asia, the farms are swallowing up rice paddies as well, and farmers are losing their livelihoods. A study in Bangladesh concluded that per unit area, shrimp farming employs only one tenth the number of people that rice growing does. Around the town of Satkhira, the shrimp industry was blamed for displacing 40 percent of the area's 300,000 people.[28]

Developments like shrimp farms and pulp plantations essentially attempt to capitalize on the intense "peaking effect" that many invasions undergo. Before it hits some limit—the collapse of the local nutrient supply, for example, or the arrival of its own pests—an exotic can sometimes divert most of the local resources into the business of growing itself. So in a very general way, a eucalyptus plantation does the same thing in fresh soil that the Leidy's comb jelly does in fresh seawater—except that there is money to be made off the eucalyptus peak. Inevitably, however, the returns will diminish, and the long-term damage is likely to be far greater than the short-term profits.

Another invasion-prone form of economic behavior

emerges from the ideology of the world trading system: "free trade" has become a global social ideal. At present, there is no comprehensive international effort to slow—or even monitor—the invasions released through trade, and the power of the free trade ideal is obviously not going to be conducive to any such efforts. In early 1998 the World Trade Organization, which adjudicates international trade disputes, was considering its first case involving an import ban based on a formal pest risk assessment. Australia has banned certain salmon imports on the grounds that they present a disease risk to native fish; the United States is calling the decision an unfair trading practice. Free trade is also invoked in U.S. timber companies' objections to the limits on raw log imports. (See also the discussion of the International Plant Protection Convention in Chapter 9.)[29]

Many protective measures could be susceptible to challenges of this sort. European nations, for example, are embargoing certain North American forest products in an attempt to prevent the spread of the oak and pine wilt pathogens. Should these diseases arrive on the continent, they could wreck many European plantations and forests. Such embargoes may grow increasingly untenable from a political point of view—even as they grow increasingly necessary from a biological one. Here, for example, are two incipient crises that may require intervention that runs counter to the current of free trade.

In the Philippines, 30–50 million palms have succumbed to a mysterious disease known as cadung-cadung, which continues to kill about a million trees a year. The disease takes years to kill—its name is Tagalog for "slowly dying"—but it is thus far completely unstoppable. Cadung-cadung has spread to Guam and perhaps to a few other islands in the Pacific. It can kill coconut palms, oil palms, and perhaps other palm species as well. If its range continues to expand, it would present a grave threat to any ecosystem in which palms are a major element, as well as to the $9-billion palm sector of the developing world's agricultural economy.

A crop disease that presents an even broader economic threat is the fatal leaf blight that emerged from the forests of Amazonia during the 1930s and wiped rubber plantations out of Brazil, where the rubber tree is native. If the blight were to reach Southeast Asia, where most of the world's rubber is now grown, then some important connections within the global economy could snap. For hundreds of applications—especially medical ones—there is no substitute for natural rubber. The rubber and palm sectors might not survive widespread contact with these pathogens, but will some future effort to prevent contagion be seen as an unfair trading practice?[30]

The threat from the global trading system is exacerbated by a consistent and nearly universal failure of individual governments to spend the money necessary to block or control invasions within their own territories, despite the scale of the losses inflicted. The U.S. federal government, for example, spends a fairly modest $150 million a year on agricultural inspection and quarantine. According to a recent audit, inspectors cannot keep up with the growing volume of incoming goods. The state of Florida is spending $25 million a year trying to chop and spray itself clear of exotic plants, and is nowhere near on top of the problem. During the past several years, for instance, Florida's hydrilla infestation has doubled in area to more than 40,000 hectares. And these are the invasions that are being fought; there are hundreds of invasions all over the world—from the Asian tiger mosquito to the Leidy's comb jelly—that are spreading largely unopposed.[31]

Yet there is compelling evidence that countermeasures make powerful economic sense. The OTA study found that every dollar spent suppressing the sea lamprey in the Great Lakes earns $30.25 in increased fisheries revenue. Preserving wetland ecosystems by eradicating purple loosestrife would return $27 for every dollar spent. And the benefits of preventing pest invasion from the imports of Siberian logs could yield a return as high as $1,661 for every dollar spent. Of course, such cost-benefit ratios are built on a great deal of conjecture—for one thing, we do not know how far we really can push exotics like the lamprey or loosestrife. But the figures show clearly that efforts to stem invasion are underfunded—not just by ecological standards, but by economic standards as well.[32]

Why aren't the economic threats of invasion sufficient to motivate the higher spending that would counter them? The obstacle here is a standard funding predicament that injures many environmental and social programs. The benefits of action, while collectively enormous, are very diffuse. Or they are only partly quantifiable, or they consist largely of damage avoided. They are not therefore likely to be visible to anyone who is not already following the issue closely. The expenses, on the other hand, are much smaller but are usually needed up front. They are visible to everyone.

Finally, there is the environmental problem that could affect just about any economic trend: the changes in global climate likely to be provoked by emissions of heat-trapping greenhouse gases, primarily carbon dioxide released from the combustion of coal and oil. During the past 130 years or so, the average annual global temperature has risen from around 13.5 degrees Celsius to around 14.4 degrees. The 1990s have been the hottest decade since recordkeeping began, and most climate scientists expect a further warming, in the range of 1–3.5 degrees by the year 2100, depending on future emissions trends and how the Earth's climate system reacts to them.[33]

That may not sound like much, but even relatively minor temperature shifts, and the consequent changes in rainfall patterns, are likely to provoke myriad new invasions or exacerbate invasions already under way. For example, the mild, wet winters and dry summers that computer climate models forecast for much of the western United States are likely to favor two of the region's worst weeds: cheat grass and Russian thistle. Longer growing seasons in the temperate zones would allow many exotic plants that are currently limited to asexual spread—for example, by sprouting from "runners"—to flower and set seed as well. This sexual reproduction would permit much more rapid adaptation and dispersal. Fast-growing, highly invasive plants like cheat may also be able to profit directly from the atmosphere's increased carbon content: in effect, they could be "fertilized" by it. (All green plants "breathe" carbon dioxide, but different species absorb carbon at different rates.) Any slower-growing natives, unable to use carbon as quickly, would tend to lose out to the invaders.[34]

The warming waters are also likely to invite additional invasions. For example, species that need warm water to breed successfully may find more room for themselves. *Caulerpa taxifolia*, the tropical seaweed that is carpeting parts of the Mediterranean, might be able to move into the North Atlantic. (See Chapter 7.) Intentionally introduced aquatic organisms could alter their behavior too. The Japanese oyster, a major aquaculture species, could become a pest in waters presently too cool to allow it to breed.[35]

Some insect populations may already be reacting to incipient climate change, since insects are often extremely sensitive to weather patterns. In Canadian boreal forest, for example, warm years often unleash huge outbreaks of spruce budworm, a native tree-chewing pest that has been expanding its range westward, apparently under the influence of a warming trend. The population dynamic of these explosions is phenomenally intense: it has been estimated that a single budworm outbreak can produce 7,200 trillion individual budworms. (That's 72 followed by 14 zeros—well over 1.2 million times the global human population.) In Costa Rica, a recent warming trend seems to have allowed the yellow fever mosquito to vault the country's central mountain range—formerly too cool for it—and occupy the western half of the country.[36]

But range expansions will not be the bugs' only response. Small insects and other invertebrates routinely conspire with the climate system to create a ready-made invasion mechanism: they are continually being swept into the air and deposited hundreds of kilometers downwind. This phenomenon has been studied a few times, usually by counting the arrivals in little plots on barren islands, and then extrapolating to larger areas. For instance, research on the Indonesian island of Krakatau (between Sumatra and Java) suggests that as many as 50 million insects are raining down on that tiny piece of volcanic real estate every day. A steady drizzle of arthropods is a normal part of the planet's meteorology. Most of the time, this insect rain has a negligible effect, because any particular area is liable to have been exposed to all the species living in the region many times over. But climate change may open up all sorts of opportunities for species that generally die on arrival.[37]

In all these ways—in its monoculture mentality, in its failure to clean up its biopollution, in its willingness to profit from the "peaking effects" of managed invasions, even in its pollution—the global economy is a homogenizing force. Natural forces, on the other hand, tend to work in the other direction—toward greater diversity. Ultimately, of course, it is the natural forces that create wealth. How much profit can there be in opposing them?

III

Remedies

9

Toward an Ecologically Literate Society

In 1942, at the height of Germany's invasion of the Soviet Union, a team of German botanists took up arms against a small exotic plant, *Impatiens parviflora*, which they believed was displacing a native relative, *I. noli-tangere*, from German forests. They did not shy away from the political angle: "As with the fight against Bolshevism, [in which] our entire occidental culture is at stake, so with the fight against this Mongolian invader, an essential element of this culture, namely, the beauty of our home forest is at stake." A year earlier, a group of landscape architects had begun drafting a "Reich Landscape Law," which would have banned the use of exotic plants in German landscapes, but Hitler's government never got around to enacting it.[1]

Exotics are a policymaker's nightmare. At first

glance, invasion biology can look like some sort of xenophobia in disguise. The situation is not improved by the fact that serious bioinvasions have actually been used to feed political paranoia—and not just among Nazi gardeners. A Soviet term for the Colorado potato beetle (see Chapter 2), was "the six-legged ambassador from Wall Street"—the idea being that the United States had deliberately introduced the pest. In the spring of 1997, Cuba was blaming the appearance of an Asian mite, a serious crop pest already widely distributed in the Caribbean, on the United States. According to Cuban officials, a crop-dusting aircraft on a mission from the U.S. Drug Enforcement Agency seeded the mite over Cuba.[2]

Because the issue carries such a heavy political burden, the policies for countering exotics must be carefully integrated into conservation policy in general. A stable political mandate to deal with exotics can only come from a public that believes not that exotics are somehow "evil," but that ecosystem integrity is worth preserving. What follows is an attempt to locate the opportunities for action on as many levels as possible—from international relations to personal actions, and from legal mechanisms to on-the-ground control techniques.

★   ★   ★   ★

Modern international agreements are a record of humanity's struggle to recognize its own common interests, and that record clearly shows a growing environmental consciousness. The first treaty on an environmental issue seems to have been the European Convention Concerning the Conservation of Birds Useful to Agriculture, signed in Paris in 1902. Today, there are around 175 environmental treaties. Some of these are as flatly utilitarian as that 1902 agreement, which aimed at preserving a valuable natural service to the economy, in the form of pest-eating birds. But other treaties are framed explicitly for the purpose of conservation. And in them, perhaps, policymakers may be gathering forces to make a historic leap in political imagination.[3]

Take, for example, the 1973 Convention on International Trade in Endangered Species of Wild Fauna and Flora (CITES), which is intended to protect endangered species from being collected for export—a major threat to many rare plants and animals. CITES was a landmark advance in conservation and it clearly makes good economic sense. Many endangered species are very valuable, which is why they were being traded in the first place, so it is obviously worth conserving them. But CITES did not grow out of this kind of utilitarian rationale; it is the product of a conservation ethic. Agreements like CITES may be the beginning of a global legal system that extends rights— at the very least, a right to existence—to other species. We have begun, however fitfully and imperfectly, to legislate globally on behalf of life in general.[4]

But that legislative effort will have to go far beyond the scope of agreements like CITES in order to deal effectively with bioinvasions. Essentially, CITES is about preventing the intentional movement of a relatively small number of clearly identified species. Fighting bioinvasions is about preventing the movement, intentional or unintentional, of many thousands of often very poorly identified species. Because of its breadth, the task constitutes a kind of challenge to prevailing assumptions that more focused concerns like CITES can avoid.

Think of it as the risk of being haunted by Nazi gardeners: that kind of xenophobia is the antithesis of just about everything contemporary environmental legislators would want to promote. And yet, we don't want exotic plants taking over forests either. We want a world in which people are as free as possible to travel and to exchange goods and ideas. But at the same time, we *need* a world in which most other living things stay put. While there is no strict logical contradiction here, there is a kind of ideological discordance. On the international level, the challenge is to use the globalization of the international legal system—the growing power of its treaties and arbitration procedures—to prevent a kind of biological globalization. And that effort has barely begun.

There is no such thing as a bioinvasion treaty, although any agreement with a conservation component could probably be construed as furnishing some basis for dealing with exotics. Of the treaties that connect with the problem on a general level, perhaps the most important is the 1992 Framework Convention on Climate Change, which aims to stabilize and eventually reduce global carbon emissions—an essential step in staving off a probable tidal wave of new invasions that could be triggered by climate change. (See Chapter 8.) Explicit coverage of invasions, in one form or another, is attempted in at least 23 global or regional agreements. (See Table 9–1.) The coverage ranges from essentially meaningless to the rigorous approach of the Antarctic treaty system, which categorically excludes all exotics from the treaty area, except for organisms listed in an annex. A complete review of all 23 agreements is far beyond the scope of this book, but three treaties are especially important.[5]

Table 9–1. Multilateral Treaties That Refer to Exotics[1]

Year	Treaty
1951	International Plant Protection Convention[2]
1958	Convention Concerning Fishing in the Waters of the Danube
1964	Agreed Measures for the Conservation of Antarctic Fauna and Flora
1968	African Convention on the Conservation of Nature and Natural Resources
1976	Convention on the Conservation of Nature in the South Pacific
1979	Convention on the Conservation of European Wildlife and Natural Habitats
1979	Convention on Migratory Species of Wild Animals
1980	Convention on the Conservation of Antarctic Marine Living Resources
1982	United Nations Convention on the Law of the Sea
1982	Benelux Convention on Nature Conservation and Landscape Protection
1985	[ASEAN Agreement on the Conservation of Nature and Natural Resources]
1991	[Protocol to the Antarctic Treaty on Environmental Protection]
1992	Convention on Biological Diversity
1992	[Convention for the Conservation of the Biodiversity and the Protection of Wilderness Areas in Central America]
1994	Agreement on the Preparation of a Tripartite Environmental Management Programme for Lake Victoria
1994	[Protocol for the Implementation of the Alpine Convention in the Field of Nature Protection and Landscape]
1994	North American Agreement on Environmental Cooperation (side agreement to the 1994 North American Free Trade Agreement)
1995	[Agreement on the Conservation of African-Eurasian Migratory Waterbirds]

[1]Does not include agreements on genetically modified organisms or human diseases; also excludes five protocols to the United Nations Environment Programme Regional Seas Conventions. Agreements in brackets are not yet in force.   [2]Included here because of its importance to the discussion, even though its conceptual basis is pests in general, rather than exotics specifically. SOURCE: See endnote 5.

The 1951 International Plant Protection Convention (IPPC, as amended most recently in 1997) is essentially a "phytosanitary agreement," a mechanism for protecting agriculture from pests that could spread through international trade in produce and other plant products. The 98 parties to the convention are supposed to maintain inspection procedures for relevant exports and to undertake eradication and control measures when new infestations occur. The convention also provides for side-agreements, which could be negotiated for particular regions, pests, or forms of transport.[6]

The IPPC has the makings of a much larger mission than crop protection: it could be extended to protect native, nonagricultural floras from pests abroad in the world trading system. This would be a "natural" extension of the power and expertise the convention has thus far assembled. The parties are already required to have national plant protection offices, for example, and side-agreement provisions could readily be adapted to deal with a broader range of pests. Extending the IPPC would recreate, on a smaller scale, that leap of the political imagination from the purely utilitarian to a concern for the stewardship of life in general.[7]

Unfortunately, however, the IPPC is at present hopping in the other direction. The recent amendments to the treaty were intended to bring it into conformity with the phytosanitary standards of the World Trade Organization (WTO). The WTO and other international trade organizations generally regard "harmonizing" standards like these as a way of promoting free trade. But harmony in this case has crippled the IPPC as an invasion-fighting tool. Under the amended treaty, sanitary measures cannot be effectively preemptive: officials in a signatory country are not supposed to take any steps that would restrict the flow of goods, in order to stop an incoming pest, until they have done a risk analysis for that particular pest. Given the thousands of exotics moving through the trading system, that is a little like deciding to fight off a military invasion by letting in the enemy soldiers and then polling each one to determine individual levels of hostility.

This approach perpetuates the antiquated notion that invasions are rare exceptions to the rule rather than the increasingly common part of the trading routine that they really are. In its current form, the IPPC looks backwards in another way as well. An organism can be defined as a pest only on the basis of economic criteria—its ecological threat is, strictly speaking, irrelevant. The WTO is not generally celebrated for its environmental sensitivities, but the amendments to the IPPC have had the remarkable effect of making the WTO's own sanitary agreement a more attractive environmental document, at least in certain respects, than the IPPC.[8]

The 1982 United Nations Convention on the Law of the Sea contains an article that requires its 125 parties to "take all measures necessary to prevent, reduce, and control pollution of the marine environment resulting from . . . the intentional or accidental introduction of species, alien or new, to a particular part of the marine environment, which may cause significant harmful changes thereto." Subsequent articles call for the development of procedures for preventing invasions and for more research on pathways and control methods. While the procedures envisioned in the treaty would be nonbinding, the basic obligation to deal with marine invasions, as set forth in Article 196, is hard-wired right into the agreement. The treaty has yet to inspire any general effort to counter marine invasions, but it could serve as a legal basis for doing so. It may help narrow the gaping ballast water pathway, for example, and it could also prove useful for regulating certain forms of marine aquaculture, such as shrimp farming.[9]

The treaty that offers the greatest opportunity is the 1992 Convention on Biological Diversity (CBD), even though its current exotics provision is a disappointment. Article 8(h) requires the treaty's 172 parties to "[p]revent the introduction of, control or eradicate those alien species which threaten ecosystems, habitats or species," but only "as far as possible and appropriate." The likely effect of that qualifying clause is to dissolve the article into an unenforceable, pro forma nod to the problem. The real potential of the CBD lies in the precursor to the current text. The initial draft of the treaty, prepared by the International Union for Conservation of Nature and Natural Resources (IUCN), would have established a scientific authority on exotics modeled on the CITES scientific committees, which determine which species merit protection under that treaty. Like CITES, the earlier CBD draft envisioned a listing process—an inventory of especially dangerous exotics that could have focused international attention on the high-priority invasions.[10]

The spirit, if not the letter, of that earlier draft could be resuscitated by establishing an expert exotics panel as a part of the CBD process. The work of the panel would be to acquaint treaty secretariats (including the CBD secretariat), national governments, and international trade agencies (such as the WTO) with the likely effects of their activities on the spread of exotics. High-priority invasions identified by the panel could be singled out for more thorough independent assessments. This process could be especially useful for reducing the biological pollution released by the world trading system. It could also be made to yield a public health dividend: by monitoring the movement of disease vectors, and in some cases of the pathogens themselves, the panel could help ground health policies in the ecology of infectious disease.[11]

★   ★   ★   ★

Treaties are intended to be binding agreements between the nations that subscribe to them, but there is another, much larger category of diplomatic literature that is nonbinding. This international "soft law" includes plans of action or intent, like Agenda 21, the famous blueprint for sustainable development that came out of the 1992 Earth Summit in Rio de Janiero. Soft law also includes a voluminous collection of codes of conduct developed by various international agencies to help regulate fisheries, the forestry sector, and many other industries. Because soft law is insulated to some degree from the intense political pressures typical of treaty negotiations, it is somewhat less likely to be compromised by political fudging. Consequently, soft law is often closer to the cutting edge of international reform than are the treaties. And soft law can sometimes "harden." Solid codes may become binding—or acquire a degree of influence just short of binding—when they are formally adopted by governments or by the international agencies themselves.[12]

An enormous amount of soft law bears on exotics, in one way or another. Agenda 21, for example, is arguably the most influential environmental document in the soft law genre, since it serves as a kind of charter for the U.N. Commission on Sustainable Development and has sparked local Agenda 21 processes in more than 1,800 municipalities in 64 countries. It is therefore unfortunate that Agenda 21 should be so weak in its coverage of exotics. Apart from a blanket recognition that exotics have sometimes caused biodiversity loss, Agenda 21 makes no attempt to outline any response to invasion, apart from suggesting that governments assess regulations on ballast water discharge. Its treatment of aquaculture and forestry is consistent with the current practice of aggressive, more or less indiscriminate deployment of exotics.[13]

But a remedy may be in the works. A kind of mini-Agenda 21, focused specifically and exclusively on bioinvasions, is now being developed under the auspices of an international research organization, the Scientific Committee on Problems of the Environment (SCOPE), in partnership with the U.N. Environment Programme, IUCN, UNESCO, and the International Institute for Biological Control. The result is intended to be a global strategy, complete with action plans and a nontechnical explanation of the need for action. Another set of guidelines, intended to complement the SCOPE strategy, has been drafted by the IUCN Species Survival Commission's Invasive Species Specialist Group (ISSG). The guidelines derive in part from the *IUCN Position Statement on the Translocation of Living Organisms,* adopted in 1987. They attempt to develop a practical framework for following through on the CBD Article 8(h).[14]

Among the codes of conduct, coverage of the issue varies as much as it does among the treaties. In its code on marine introductions, for example, the North Atlantic regional science organization, the International Council for the Exploration of the Sea

(ICES), requests that when any of its 19 member states are considering a marine introduction, they submit a detailed plan that would in effect include an environmental impact assessment. On the other hand, the aquaculture section of the fisheries code published by the U.N. Food and Agriculture Organization (FAO) contains just three relevant provisions: one urges international consultation for introductions into transboundary waters; another says that "efforts should be undertaken to minimize the harmful effects" of exotics; and the third is a vague appeal to "international codes of practice." These provisions fail by the standards the FAO itself set a year later, with the publication of its *Precautionary Approach to Capture Fisheries and Species Introductions*—a set of guidelines that relies heavily on the ICES code.[15]

Effective codes must usually descend into the nuts and bolts of an issue in order to make recommendations that really mean anything. But there are a few broad principles that could be used to guide the process of reforming or implementing codes, and soft law in general. The following six steps could form the core of an agenda for countering accidental releases.

- Locate the responsible parties. Designate national and international authorities for monitoring and responding to accidental releases, as the IPPC has done for crop pests. Identify the industries and agents that are doing the polluting, if these are not already obvious, and set up clearly defined alliances with them.
- Insist on reasonable precautions. Where sound procedures have already been developed for narrowing a pathway, these should be adopted.
- Tackle the problems that already exist. Develop

control and eradication procedures for known pests, and look for ways to narrow the pathways through which they are moving.

- Develop emergency response capabilities. Emergency protocols should include an early warning system, a means of contacting relevant experts on short notice, and a pool of readily available funds. Exotics are often much more vulnerable in the initial phase of an invasion than they are after they have had a chance to establish themselves.

- Develop a capacity for analysis. When a new problem arises, it should be possible to follow it back to its causes.

- Ask the polluter to pay. These efforts should be supported in large measure by the industries that are doing the polluting. It is not practical or reasonable to demand that industries shoulder the burden alone, since bioinvasions are such a broad social problem. But substantial industry participation should be a part of good corporate citizenship—and something that individual citizens of all countries have a right to expect.

Three additional principles would apply just to intentional introductions:

- Assume it is dangerous unless there is strong evidence to the contrary. This is the bioinvasion version of the "precautionary principle," an important mechanism for shifting the burden of proof off of those who are arguing for conservation and onto those who are arguing for some form of exploitation.

- Require an environmental impact assessment, as with the ICES code. The assessment should include a mechanism for all interested parties—

local people, industries, neighboring states, and others—to air any concerns they may have.
- Make a serious effort to encourage the use of native species. Some natural resource–based industries pursue a nearly indiscriminate use of exotics without attempting any evaluation of local biological resources. Thousands of tropical tree species, for example, have yet to be evaluated for their silvicultural potential. In North America, there are 4,000–5,000 native bee species with some pollination ability; a native pollinator industry could help ease agriculture's honeybee famine and promote the conservation of native insects.[16]

★   ★   ★   ★

Information on exotics is badly fragmented—it is scattered about in hundreds of technical newsletters and publications, in professional discussion groups, and among small cadres of regulators. That makes it very difficult for outsiders to get their bearings in the field, or for insiders to keep themselves current. On any level of activity, from international negotiations to the management of a particular natural area, fragmentation of the knowledge base can be an obstacle to progress. So a highly efficient way to catalyze progress would be to develop a comprehensive, readily accessible, global pool of information on exotics. This might be done as part of the SCOPE initiative, which envisions an "Information Clearinghouse," or within IUCN's ISSG.[17]

A peer-reviewed journal devoted exclusively to bioinvasions is now in the planning stages. The journal could serve as the foundation for a kind of virtual library on the subject, a world information base on bioinvasions. The preliminary agenda for such an effort might include the following three items.

- Constructing and maintaining a master linking page to the hundreds—perhaps thousands—of on-line list servers and Web pages that deal with various aspects of bioinvasion.[18]
- Developing an on-line invasive species database, of the sort the ISSG apparently envisions. The database could include basic facts on the ecology of the organisms listed, a list of known invasions, a bibliography, and references to relevant experts and other databases.[19]
- Monitoring the effect of international law on the spread of exotics. This effort might involve regular reporting on treaty negotiations, litigation, and so forth. It could also develop a kind of "legal memory" for the field—an archive of laws and regulations pertaining to exotics and a record of relevant national and international litigation. (Perhaps this effort could be undertaken in conjunction with the IUCN Environmental Law Centre.)

*   *   *   *

Bioinvasions are tightly bound up with local economies and cultures, which of course vary greatly from one place to the next. In many countries, forging a coherent national policy on exotics is likely to be extremely difficult, since the relevant concepts themselves may be "exotic" to the local cultural or legal terrain. But general policy objectives could be derived from the code of conduct principles outlined above, and a possible model for comprehensive national legislation comes from New Zealand.

In 1993, New Zealand passed its Biosecurity Act, which collected earlier legislation on pest management and exclusion into a single, coherent, substantially expanded law. In effect, the Biosecurity Act became a blanket authority for dealing with virtually all forms of biological pollution. In 1996, the country passed a complementary piece of legislation, the Hazardous Substances and New Organisms Act (HSNOA), which establishes a single agency, known as the Environmental Risk Management Authority, for assessing all applications to import or manufacture potentially dangerous substances or organisms.[20]

In early 1998, regulations were still being written for the HSNOA, so it is not yet possible to gauge its actual performance. But several features of the New Zealand approach are strongly encouraging. For example, a clear intent of both acts is to integrate all the responsible authorities and regulations. That might seem like too obvious a legislative requirement to mention, but policy on exotics is often deeply confused and dispersed, which can lead to all sorts of administrative contradictions. In the United States, for example, some officials in the Michigan Department of Natural Resources have been attempting to rid the state of certain invasive exotic plants that other parts of the same Department are still planting. That is an acute form of a common problem: one land management agency promoting exotic organisms that another is desperate to eradicate.[21]

Another strength of the HSNOA is its use of the precautionary principle: any "new" organism is in effect considered dangerous and its potential environmental impact must be reviewed before it can legally be brought into the country. (A "new" organism for the purposes of the act is any organism not already present in New Zealand, including genetically engineered organisms that might be developed within the country and exotics already eradicated from it.) The HSNOA is concerned with intentional introductions, of course, but it is still instructive to see how far this approach has moved from the pest-fighting standards of the IPPC, which basically assumes that exotics are safe until proved otherwise. New Zealand's legislators, even those only casually familiar with the natural history of their country, are well positioned to see the problems with that assumption. (See Chapter 5.) Today, some sort of formal, public environmental impact assessment for new introductions is essential for the preservation of any country's natural wealth.

The analogy between chemical and biological pollution is useful for thinking through how such assessments should work. The right to manufacture or import a chemical does not necessarily exempt an agent from liability for misuse or unforeseen effects. In a similar way, an approved introduction should not confer the right to release an organism indiscriminately. Approval should be contingent upon adherence to professional codes, where these exist. It might also be useful to make introductions subject to very long statutes of limitation, since it may take years for an exotic to become invasive.[22]

But there's a point at which the analogy with chemical pollution breaks down. Chemicals are inert; living things are active—they multiply and adapt. So an uncomfortable element of conjecture haunts even the most carefully reviewed intentional introductions. And dealing with the accidental releases is like fighting a guerilla war. One characteristic of many highly successful invaders, for example, is that they are "polyvectic"— they exploit many different pathways simultaneously. Take the tiny Argentine ant: it travels on or in just about anything that moves—plants, animal fodder, military equipment—and has become a major ecological headache on every continent except Antarctica and on many island groups. (Most of the disruption results from the displacement of native insects; see the discussion of the Cape Floral Kingdom in Chapter 5 for an example.) The green crab is polyvectic too. (See Chapter 7.)[23]

Simple, passive leak-plugging will not work with creatures like these. The policies must be as active as the targets; they must be designed to follow the invaders back into the complicated, shifting machinery of trade and industry whence they came. An example of what can be accomplished in this regard is the U.S. National Invasive Species Control Act of 1996, which focuses on aquatic invasions. The act includes a pilot program in which all ships bound for U.S. ports are supposed to pump out and refill their ballast tanks before entering U.S. waters, or use some alternative ballast water treatment, assuming one is available. Because most ballast exotics are coastal organisms, relatively few can survive in full-strength seawater. But this ballast water exchange technique is not completely effective, and in rough seas it can be dangerous, so the act also includes incentives for developing those alternatives. At present, the best options appear to be more sophisticated filters, heating, electrical pulses, and biocides (clearly not the preferred alternative from a broader environmental perspective). The act also funds additional research into invasions themselves.[24]

It is essential to look for the outward leaks as well as the inward ones. In 1988, the U.S. government reacted to the Asian tiger mosquito invasion by imposing disinfection requirements on tire imports—but not on tire exports. That left the mosquito with a well-established pathway into other countries. And what goes around comes around: mosquitoes going out in used tires could very well mean more dengue fever coming back in the bloodstreams of U.S. tourists. Given the comprehensiveness and volume of global trade and transportation, every nation's biosecurity has become every other nation's biosecurity.[25]

*   *   *   *

Within natural resource programs, it may be possible to end certain categories of exotic deployment at public expense—the use of exotic forage plants on rangeland, for instance, or of exotic grasses for soil conservation. As late as 1995, retired croplands in the U.S. Conservation Reserve Program were 2.5 times more likely to be planted in exotic grasses than native ones, at least partly because native grass seed has been hard to come by.[26]

But consistent movement away from exotics should stimulate interest in native replacements. The state of Illinois, for example, now excludes virtually all exotics from its land management programs—its natural areas conservation, roadside plantings, and so on. That has in effect turned the state nursery operation into a kind of engine for ecological restoration: the nurseries produce native plants—grown mostly from diverse, wild-collected seed—and the region's natural flora are getting a second chance.[27]

Sustainable agriculture offers similar opportunities, especially in developing countries. There are the local landraces (traditional varieties), of course, and in many places there are entire crop species that once were grown and are now largely forgotten. Since these species are usually native, they are generally better adapted to local conditions than standard commercial crops. Sub-Saharan Africa alone has an estimated 2,000 native food plants, including native grains, vegetables, root crops, and fruit. In a similar way, native species initiatives could be developed in agroforestry, forestry, and aquaculture.[28]

In many places, the managers of natural resource programs may be uniquely qualified to help people, especially young people, develop a sense of natural place—an understanding of how and why one region differs from another. Until "that tree" becomes an Ailanthus or a black gum, until "those birds" become house sparrows or gold finches, conservation will not acquire the kind of grip on the public imagination that it will need if it is going to succeed.

A solid mandate for dealing with exotics will require this kind of general ecological literacy. After all, exotics can be beautiful, and landscape restoration has its uglier moments. Chopping out invasive trees to restore native North American prairie, or poisoning a California lake to rid it of the aggressive exotic northern pike—such efforts may be technically sound but they look like a huge insult to nature. People will see the destruction readily enough, and they can hardly be blamed for reacting to it if they do not understand the healing that follows.[29]

A degree of ecological literacy should also make it easier for people to accept a fact that may be obvious to ecologists, but is likely to seem counterintuitive in the context of our general consumer culture. The idea is simple enough: nature imposes real, permanent limits on what people can do without provoking some sort of ecological collapse. There may be no inherent natural limit to economic growth, if "growth" is construed simply as an increase in the value of available goods and services. But there is certainly a natural limit to the amount of grain that we can grow, or the amount of fresh water we can safely use. In a similar way, there are limits to the amount and type of biotic mixing that an ecosystem will tolerate. It may, for example, never be possible to import Siberian raw logs safely into western North America. If you live in that region and you value it, the prospect of making do without Siberian logs and the forest pests within them probably will not break your heart. If you don't care about the forests, or if the idea of never being able to do something seems strange, then that is a measure of how far we still have to go— even in our thinking.

*　　*　　*　　*

Within the limits of our current technology and ecological understanding, we have to assume that invasions are generally permanent. But their effects can still be greatly reduced, and total eradication of an established exotic is sometimes possible. Either way, control or eradication of exotic species is now as important a task for conservation as saving endangered ones— indeed, there may be little point in attempting the latter task without also undertaking the former.[30]

Exotics can make a mess, not just of the areas they invade, but of conservation principles as well. However distasteful the idea of using pesticides in natural areas may be, a certain amount of chemical pollution is sometimes essential for controlling the biological pollution. Invasive woody plants, like the melaleuca in south Florida (see Chapter 1), are often attacked by "hack and squirt" crews, who cut away the offending brush and apply a herbicide to the stumps to prevent resprouting.[31]

Fortunately, there is usually a great deal more hacking than squirting. One of the most ambitious efforts in this regard is South Africa's "Working for Water" program, a government-organized consortium of municipal authorities, environmental and social nongovernmental groups, companies, universities, and other organizations. Drawing on both public- and private-sector funding, the program is restoring water tables by clearing exotic trees, such as pine and acacia. So far, more than 33,000 hectares have been cleared, freeing up as much as 3,500 cubic meters of water per hectare per year in densely infested areas. The effort, begun in 1995, has provided jobs to more than 6,600 people, over half of them women.[32]

Direct, carefully targeted attacks can work on animal exotics as well—at least on vertebrates. Rats have been successfully eradicated from more than 60 small islands off the coast of New Zealand and from an island in the Antiguan group in the Caribbean. Poisoned bait was the technique used, and success was due in large measure to the fact that the island rats had become in their way as "naive" as native island wildlife. (See Chapter 5.) Rats are clever, social animals. In a challenging environment, they may take days to explore a new food source, and the fate of the more adventurous diners may be noted by the rest. That is one of the reasons rats will never be poisoned out of places like New York City. But the island rat "cultures" had never encountered poison before and took no precautions.[33]

A cleverer if uglier technique has been used successfully against another dangerous island invader. Aldabra Atoll, a World Heritage Site within the Indian Ocean Republic of the Seychelles, is being cleared of goats as part of a World Bank environmental action plan. The idea is to shoot all the goats, but the survivors of the initial campaign grow extremely wary, so the project uses a "Judas goat." A goat is fitted with a radio tracking collar and released onto the island. Goats dislike solitude and are very good at finding other goats. The Judas goat's impulse to seek the company of its fellows betrays their whereabouts to the hunters.[34]

Unless you know what is at stake, it may be difficult to work up much enthusiasm for poisoning and shooting. But programs such as these are critical for saving many rare plants and animals. In the case of the goats, it is hardly an exaggeration to say that their removal is the key to saving entire island ecosystems. On Aldabra, as on many islands, the presence of goats can mean near total destruction of low-level vegetation, and eventually the animals that depend on it. The loss of plant life on Aldabra, for instance, threatens the atoll's endangered giant tortoises and birds.[35]

But many exotics multiply too quickly over too large an area to be susceptible to direct attack. When it comes to creatures like European starlings, zebra mussels, purple loosestrife, or mosquitoes, people cannot usually kill enough of them to make any difference. Some of these invaders can be dealt with by recruiting other organisms to do the killing instead—a technique known as biological control or biocontrol. As usually practiced today, biocontrol attempts to restore a natural balance by releasing one or more of the offending exotic's native pests into the area it has invaded. Effective biocontrol agents for an invasive plant, for example, might be found among the insects that feed on it in its native range. Biocontrol will not eradicate an invader, but it can sometimes reduce it to the point of ecological insignificance. Both the invader and the agent would remain in the environment, but at very low levels. Any increase in the invader population would be answered by an increase in the agent population, which would force it back down.

All sorts of activities have been pulled under the rubric of biocontrol, many of which no informed advocate of the technique would endorse today. During the nineteenth and early twentieth centuries, for example, a crude form of biocontrol attempted to suppress island rat populations by introducing exotic predators like the mongoose. Mongoose populations on many islands are a sad legacy of such experiments. (See Chapter 5.)

Some of these attempts took on a clumsy, falling-upstairs sort of tempo, in which each step triggered a response more damaging than the last. Take the saga of rat control on the islands of Micronesia. The effort began by importing giant monitor lizards. But rats are active at night; they are not suitable prey for the monitors, which are active during the day. So the monitors turned to poultry instead. Sometime before 1945, an enormous South American toad, called the cane toad, was imported to give the monitors something else to eat and perhaps also to keep down insects in the coconut plantations. (The cane toad gets its common name from the fact that it has been released all over the world's sugar-growing regions in the hope that it would control insects and even rats—a task it has consistently refused to perform.)[36]

The toad secretes a powerful venom from its skin, and large numbers of monitors were poisoned. As the monitors died off, one of the coconut pests, a rhinoceros beetle, underwent a population explosion because the monitors had been eating its grubs. With the monitors out of the way, the toad population exploded too.

Cats, dogs, and pigs attacked the toads and were killed in their turn. Then the rat population exploded because the cats and dogs had been preying on the rats. The giant African snail, brought in by the Japanese during World War II as a food, exploded as well, perhaps partly because of all the available carrion in the form of cat and dog carcasses. During the 1970s and 1980s, a predatory flatworm was introduced to try to control the snail. The flatworm is currently spreading throughout the islands and has become a major new threat to Oceania's extraordinarily diverse native snail fauna.[37]

Today, the idea of using mongooses to control rats would garner about as much scientific support as using anthrax to control overgrazing. And yet this kind of debacle is not entirely a thing of the past. It continues in aquatic form with the release of mosquito fish species all over the world for mosquito control. Mosquito fish are voracious predators of mosquito larvae, but they will also eat many other tiny creatures. As soon as the supply of mosquito larvae begins to diminish, the menu broadens—and there are always plenty of left-over mosquitoes. One mosquito control expert, pondering his colleagues' continued enthusiasm for these introductions, recently asked the obvious question: if mosquito fish are so effective, "how is it there are so many mosquitoes in areas that are its native habitat?" Exotic mosquito fish have caused the extinction of native fish by eating their larva and fry.[38]

But biocontrol of a very different sort is one of the best weapons available for countering many exotic plants, insects, and other invertebrates. The key to making the technique work safely is "host specificity"— the ideal biocontrol agent preys exclusively on the target organism and pursues it to the edge of oblivion.

That is why introducing any vertebrate for biocontrol is asking for trouble. Vertebrates are just too adaptable; they cannot be trusted to die off cooperatively when their favorite food gets scarce. Even among the invertebrates, biocontrol candidates must be carefully screened. Presumably, thoughtful testing would have easily detected that flatworm's ability to eat just about any snail it encounters. And better testing would presumably have prevented the release of parasitic wasps as biocontrol agents for certain crop pests in Hawaii. The wasps apparently caused the extinction of some native moths and may be a major factor in the general decline of the native Hawaiian insect fauna.[39]

And yet, among the insects and their relatives, it is possible to find creatures so exquisitely adapted to preying on only a single host that they seem almost to be a part of that organism. There are weevils, for example, whose entire existence is built around the buds of a single species of plant. Detach the weevil from the bud and the only thing the weevil can do is die. The bud is its only possible landscape, and therein lies its extraordinary power for restoring landscapes that its host has devoured. Take the invasion of the South American water fern, *Salvinia molesta*, into Papua New Guinea's Sepik River. By 1980, a decade after it first appeared, the fern had made the Sepik unfishable and impassable, starving out local villages in the process. By 1990, the infestation had been largely cleared by the introduction of a weevil, one of a group of insects that regulates the fern's growth in its native Brazil.[40]

A recent biocontrol victory of enormous social significance was the control of the cassava green mite across the entire midsection of Africa. Cassava is a root crop from the New World tropics; it grows on some of

Africa's poorest soils and feeds some of the continent's poorest people. For some 200 million inhabitants of the vast "cassava belt," an area one and a half times the size of the United States, cassava is the primary staple. That is why the arrival of the green mite, one of the cassava's native predators, was a social disaster in the making. By 1980, some 15 years after the mite was first noticed in a Ugandan field, it had spread throughout the belt, cutting yields by 50 percent or more. After a decade-long search for suitable biocontrol agents, an international team of scientists released another mite—a tiny but ferocious native predator of the green mite—in a field in Benin in 1993. It too has now spread throughout the belt, and cassava yields have thus far improved by 30 percent on average.[41]

Microbes have occasionally been recruited into biocontrol efforts as well. An epidemic can knock a pest population down quickly, but over the long term the pest is likely to develop substantial immunity. Australia's continual war with the European rabbit has followed this dynamic. In the early 1950s, a Brazilian rabbit virus was introduced into the country and the resulting plague broke the rabbit's grip on Australia's rangeland. But the virus's punch eventually weakened. By the mid-1990s, the rabbit was doing well over $100 million worth of damage a year. In October 1995, another rabbit virus escaped from a quarantined island off Australia's southeast coast, where it was being tested as a biocontrol agent. (This virus was discovered in China in 1984 and has since spread to Europe and Mexico.)[42]

Hoping to make the most out of the new virus's initial wallop, the government went ahead with a full release in the following year. Thus far, the strategy seems to have worked. Rabbit populations have crashed throughout the southern part of the country. Native plants that disappeared from some areas decades ago are returning, and the flush of vegetation is benefiting kangaroos and other native herbivores. Australia has not given up on the Brazilian virus; the fleas that transmit that virus do not do well in very dry conditions, so a new species of arid-land flea has been introduced as an additional vector. Another of Australia's pests may also be targeted for infection. Scientists are studying the possibility of releasing a pathogen of the cane toad, to try to stem that creature's advance into natural areas in the north of the country. But in most countries, concerns that microbes could attack nontarget organisms have greatly limited their use as biocontrol agents thus far.[43]

Biocontrol will never be a panacea: many factors may be controlling the population of an organism in its native range and there may not be an effective biocontrol agent among them. Even when a likely candidate is identified and released, it may not be able to achieve control on its own. Estimates of the technique's overall success rate range around 10–20 percent, although biocontrol advocates argue that the rate would improve substantially if biocontrol were not so often the choice of last resort. Biocontrol agents have been derived from the natural wealth of at least 98 countries; at least 121 countries have had biocontrol agents released into them. Many of the world's worst invaders—purple loosestrife and water hyacinth, for example—are good candidates for this technique, either alone or in combination with other strategies.[44]

The technical arsenal for dealing with exotics is relatively limited, but current research may be on the verge of expanding it. Some approaches aim to fuse biocontrol with biotechnology, in effect attempting to take gene splicing to a landscape level. It is now possible, for example, to genetically engineer viruses that cause sterility in rabbits, mice, and presumably other mammals. Whether it will be safe to release such organisms, of course, is another matter entirely. Scientists have also recently discovered a genetic technique that allows for the mass production of male Mediterranean fruit flies—that is important because it is likely to boost the power of "sterile male" releases against these and perhaps other invasive insects. (See the discussion of the lamprey in Chapter 4.) More distant and far less certain possibilities include releasing genetically engineered mosquitoes to spread disease-fighting traits into the world's mosquito populations through interbreeding. One project, for example, envisions the spread of a gene to block replication of the dengue fever virus; another would have mosquitoes producing malaria antibodies, which they would inject into humans when they bite.[45]

There are several promising developments outside biotech—for example, marine biocontrol. One potential target is the green crab, which could be attacked with a weird barnacle that is one of the crab's native parasites. The barnacle replaces the crab's gonads with itself, thereby rendering its host sterile. There are now also contraceptive "vaccines" for some mammals, and artificially produced pheromones (signaling chemicals) that confuse the mating instincts in certain insects. Given the rate at which invasions are proceeding, research on these and other "counterinvasion" techniques deserves a high priority on the conservation agenda.[46]

★   ★   ★   ★

The greatest cause for hope, however, is not to be found in some laboratory or treaty negotiation. It's in you—in your ability to relate to landscape or seascape, to whatever locale is significant to you. What lives in the green area nearest your home?

I live on the outskirts of Washington, D.C. Four hundred years ago, my area was on the fringes of what was then the most diverse temperate-zone forest in the world (in terms of the number of canopy tree species). Now most of it is a scab of asphalt, shopping malls, and tracts of suburban houses, like the one my family owns. Sometimes I go walking through the sliver of remnant forest that lines the stream below my property. There I find American beech, black willow, green ash, and black cherry—all native to my area. But at the base of my lawn there's a thicket of Ailanthus, from China— "stink tree," as it is sometimes aptly known. I see mimosa, a small weedy tree from central Asia, growing here and there as well. Much of the ground is covered by Japanese honeysuckle, an invasive exotic vine. An even more aggressive Asian vine, mile-a-minute weed, invaded my area several years ago. In the summers, it can grow a good 15 centimeters in a single day and it's covered with thorns. In the winter, it dies back, leaving a thick skein of dead stems hanging from lower branches and undergrowth.

The shrub multiflora rose, from Japan, dominates many of the clearings, except where it's overgrown and crushed by the exotic vines. I see the odd clump of privet, also from East Asia, and English ivy. In deep shade where the vines thin out, Eurasian garlic mustard covers the ground. Farther upstream, a clone of Asian bamboo is spreading. Every few years, the cherries and willows are thick with gypsy moth. Starlings and house sparrows (from Europe) and pigeons (from North Africa) are the most common birds where I live. Four hundred years ago, my area was home to one of the Chesapeake region's many Algonquian-speaking peoples: the Tauxenent. If I could summon a few Tauxenent to walk with me today, "nature" would look almost as alien to them as the shopping malls.[47]

Conservation offers a huge opportunity for amateur naturalists, and it's far too important to be left to the professionals. The best defense any area can have is a group of people who know it well, who care about it, who have the patience to try to understand what is happening in it. Conservation works best as a civic activity—the acclimatizers, for all their ecological naiveté, had that right.

Most landscapes offer a spectrum of opportunities for direct personal action. Many places, for example, are probably seen by hundreds of people but watched by almost no one. Simply recording what is happening in them could invest them with a far greater public value. ("There are rare plants growing in our neighborhood? Rare birds nest there?") Conservation organizations and even some government agencies need volunteer labor to do the hands-on work of conservation—from maintaining trails to clearing areas of invading plants. And there may be opportunities to act as an advocate for the landscape, before planning commissions, local industry groups, and other forums that influence local development.

Much of this may sound like pretty basic conservation work, rather than a focused plan for "counterinvasion." But that's exactly the point: in the long run, the only real hope against invasion is a public that values the creatures that belong where they are.

Notes

CHAPTER 1. Evolution in Reverse

1. Carroll Lane Fenton and Mildred Adams Fenton, *The Fossil Book: A Record of Prehistoric Life* (New York: Doubleday, 1989).
2. Cases of bioinvasion mentioned throughout this chapter that are not specifically referenced are documented in Chapters 2–7.
3. E.C. Pielou, *After the Ice Age: The Return of Life to Glaciated North America* (Chicago: University of Chicago Press, 1991).
4. Bioinvasion as second to habitat loss from Don C. Schmitz and Daniel Simberloff, "Biological Invasions: A Growing Threat," *Issues in Science and Technology,* summer 1997; melaleuca from Don C. Schmitz et al., "The Ecological Impact of Nonindigenous Plants," in Daniel Simberloff, Don C. Schmitz, and Tom C. Brown, eds., *Strangers in Paradise: Impact and Management of Nonindigenous Species in Florida* (Washington, DC: Island Press, 1997); native plants from Craig Diamond, Darrell Davis, and Don C. Schmitz, "Economic Impact Statement: The Addition of *Melaleuca quinquenervia* to the Florida Prohibited Aquatic Plant List," in Ted D. Center et al.,

eds., *Proceedings of the Symposium on Exotic Pest Plants* (Washington, DC: U.S. Department of the Interior, National Park Service, 1991).

5. Reginald R. Reisenbichler, "Genetic Factors Contributing to Declines of Anadromous Salmonids in the Pacific Northwest," in Deanna J. Stouder et al., eds., *Pacific Salmon and Their Ecosystems* (London: Chapman and Hall, 1997).

6. Roger G. Kennedy, *Hidden Cities: The Discovery and Loss of Ancient North American Civilization* (New York: Free Press, 1994); Alfred W. Crosby, *Ecological Imperialism: The Biological Expansion of Europe, 900–1900* (Cambridge, U.K.: Cambridge University Press, 1986); Sheryl Barbic, "Brazil's 'Genocide Decree'," *Multinational Monitor*, March 1996.

7. The "tens rule" is described in Mark Williamson, *Biological Invasions* (London: Chapman and Hall, 1996). Williamson's book is probably the best recent survey of invasion biology and certainly the most thorough inquiry into the degree to which actual invasions can be "captured" in theory.

8. On the importance of diversity for stability, see, for example, Jill McGrady-Steed, Patricia M. Harris, and Peter J. Morin, "Biodiversity Regulates Ecosystem Predictability," *Nature*, 13 November 1997.

9. Forty years of study based on an inception of invasion biology in the year when the field's first comprehensive work appeared: Charles Elton, *The Ecology of Invasions by Animals and Plants* (London: Methuen, 1958).

10. Wide and narrow range example from Edward L. Mills et al., "Exotic Species and the Integrity of the Great Lakes," *BioScience*, November 1994; melaleuca and catclaw mimosa from Quentin C.B. Cronk and Janice L. Fuller, *Plant Invaders: The Threat to Natural Ecosystems*, WWF and UNESCO "People and Plants" Conservation Manual 2 (London: Chapman and Hall, 1995), from I.D. Cowie and P.A. Werner, "Alien Plant Species Invasive in Kakadu National Park, Tropical Northern Australia," *Biological Conservation*, vol. 63 (1993), pp. 127–35, and from Williamson, op. cit. note 7; pond-apple from Faith Campbell, exotic species expert, Western Ancient Forests Campaign, discussion with author, April 1998.

11. Francis G. Howarth, Gordon Nishida, and Adam Asquith, "Insects of Hawaii," in Edward T. LaRoe et al., eds., *Our Living Resources: A Report to the Nation on the Distribution, Abundance, and Health of U.S. Plants, Animals, and Ecosystems* (Washington, DC: U.S. Department of the Interior, National Biological Service, 1995); R.H. Morman, D.W. Cuddy, and P.C. Rugen, "Factors Influencing the Distribution of Sea Lamprey

(Petromyzon marinus) in the Great Lakes," *Canadian Journal of Fisheries and Aquatic Sciences,* November 1980.

12. Jeff Crooks and Michael E. Soulé, "Lag Times in Population Explosions of Invasive Species: Causes and Implications," and Vandana Shiva, "Species Invasions and the Displacement of Cultural and Biological Diversity," both in Odd Terje Sandlund, Peter Johan Schei, and Åslaug Viken, eds., *Proceedings of the Norway/UN Conference on Alien Species* (Trondheim: Directorate for Nature Management and Norwegian Institute for Nature Research, 1996); lag time until discovery from U.S. Congress, Office of Technology Assessment, *Harmful Nonindigenous Species in the United States* (Washington, DC: September 1993); Figure 1–1 from Victoria Nuzzo, "Distribution and Spread of the Invasive Biennial *Alliaria petiolata* (Garlic Mustard) in North America," in Bill N. McKnight, *Biological Pollution: The Control and Impact of Invasive Exotic Species* (Indianapolis, IN: Indiana Academy of Science, 1993) (topographic quadrangles are approximately 590 square kilometers in area).

13. Edward Tenner, *Why Things Bite Back: Technology and the Revenge of Unintended Consequences* (New York: Alfred A. Knopf, 1996); Williamson, op. cit. note 7.

14. Curtis C. Daehler and Doria R. Gordon, "To Introduce or Not To Introduce: TradeOffs of Nonindigenous Organisms," *TREE* (Trends in Ecology and Evolution), November 1997, gives the standard formulation: "The strongest predictor of negative impacts of a nonindigenous organism remains whether it has had negative impacts in other areas to which it has been introduced."

15. Inability of time to heal invasions from Bruce Coblentz, "Exotic Organisms: A Dilemma for Conservation Biology," *Conservation Biology,* September 1990.

CHAPTER 2. The Fields

1. Bruce D. Smith, *The Emergence of Agriculture* (New York: Scientific American Library, 1995).

2. Ibid.; Tim Dyson, *Population and Food: Global Trends and Future Prospects* (London: Routledge, 1996); U.S. Department of Agriculture (USDA), Foreign Agricultural Service, *World Agricultural Production* (Washington, DC: August 1997).

3. Marilyn D. Fox, "Mediterranean Weeds: Exchanges of Invasive Plants between the Five Mediterranean Regions of the World," in F. di Castri, A.J. Hansen, and M. Debussche, eds., *Biological*

Invasions in Europe and the Mediterranean Basin (Boston, MA: Kluwer Academic Publishers, 1990); H.N. le Houérou, "Plant Invasions in the Rangelands of the Isoclimatic Mediterranean Zone," in R.H. Groves and F. di Castri, eds., *Biogeography of Mediterranean Invasions* (Cambridge, U.K.: Cambridge University Press, 1991); Richard N. Mack, "Temperate Grasslands Vulnerable to Plant Invasions: Characteristics and Consequences," in J.A. Drake et al., eds., *Biological Invasions: A Global Perspective,* SCOPE 37 (Chichester, U.K.: Wiley, 1989); Figure 2–1 based on R.H. Groves, "The Biogeography of Mediterranean Plant Invasions," in Groves and di Castri, op. cit. this note, and on Peter M. Vitousek et al., "Biological Invasions as Global Environmental Change," *American Scientist,* September–October 1996.

4. Carla M. D'Antonio and Peter M. Vitousek, "Biological Invasions by Exotic Grasses, the Grass/Fire Cycle, and Global Change," *Annual Review of Ecology and Systematics,* vol. 23 (1992).

5. W.D. Billings, *"Bromus tectorum,* a Biotic Cause of Ecosystem Impoverishment in the Great Basin," in G.M. Woodwell, ed., *The Earth in Transition: Patterns and Processes of Biotic Impoverishment* (Cambridge, U.K.: Cambridge University Press, 1990); Richard N. Mack, "Temperate Grasslands Vulnerable to Plant Invasions: Characteristics and Consequences," in Drake et al., op. cit. note 3; Faith Thompson Campbell, "Exotic Pest Plant Councils: Cooperating to Assess and Control Invasive Non-Indigenous Plant Species," in James Luken and John Thieret, *Assessment and Management of Plant Invasions* (New York: Springer-Verlag, 1997); Faith Thompson Campbell, "How Invasive Plants Destroy the Environment— New England and the World," *Newsletter of the Connecticut Botanical Society,* winter 1996; Lee Otteni, "Department of the Interior Proposes Partnership Plan for Management of Exotic Weeds," *Restoration and Management Notes,* winter 1996; fire-adapted grasses from D'Antonio and Vitousek, op. cit. note 4.

6. Ancient rat traps from A.B. Lazarus, "Progress in Rodent Control and Strategies for the Future," in R.J. Putman, ed., *Mammals as Pests* (London: Chapman and Hall, 1989); rat origins from Christopher Lever, *Naturalized Mammals of the World* (London: Longman, 1985); Paul Epstein, "The Threatened Plague," *People and the Planet,* vol. 6, no. 3 (1997).

7. William B. Showers, "Diversity and Variation of European Corn Borer Populations," in Ke Chung Kim and Bruce A. McPheron, eds., *Evolution of Insect Pests: Patterns of Variation* (New York: John Wiley, 1993); Jenny Luesby, "High-tech Corn

Seeds Start US Patents Battle," *Financial Times,* 22 March 1996.

8. Gordon G. Whitney, *From Coastal Wilderness to Fruited Plain: A History of Environmental Change in Temperate North America 1500–Present* (Cambridge, U.K.: Cambridge University Press, 1994).

9. Alfred W. Crosby, *Ecological Imperialism: The Biological Expansion of Europe, 900–1900* (Cambridge, U.K.: Cambridge University Press, 1986).

10. Ibid.; Lever, op. cit. note 6.

11. Bison ecology from Jane H. Bock and Carl E. Bock, "The Challenges of Grassland Conservation," in Anthony Joern and Kathleen H. Keeler, eds., *The Changing Prairie: North American Grasslands* (New York: Oxford University Press, 1995); reduced floral diversity from Whitney, op. cit. note 8. The intermountain region lies between the Rocky Mountains and the Cascade–Sierra Nevada Ranges.

12. Overgrazing and the cheat invasion from Campbell, "Exotic Pest Plant Councils," op. cit. note 5; cheat as spring forage from Robert Devine, "The Cheatgrass Problem," *The Atlantic,* May 1993; cheat root system from le Houérou, op. cit. note 3.

13. Whitney, op. cit. note 8; Mack, op. cit. note 5.

14. "Legumes: Glycine wightii," <http://ifs.plants.ox.ac.uk/fao/tropfeed/data/r243.htm>, viewed April 1998; kudzu from U.S. Congress, Office of Technology Assessment (OTA), *Harmful Nonindigenous Species in the United States* (Washington, DC: September 1993).

15. Andy Dobson, "The Ecology and Epidemiology of Rinderpest Virus in Serengeti and Ngorongoro Conservation Area," in A.R.E. Sinclair and Peter Arcese, eds., *Serengeti II: Dynamics, Management and Conservation of an Ecosystem* (Chicago: University of Chicago Press, 1995); Masai quote from Herbert H.T. Prins and Henk P. van der Jeugd, "Herbivore Population Crashes and Woodland Structure in East Africa," *Journal of Ecology,* vol. 81 (1993).

16. Epidemics from Dobson, op. cit. note 15; Hamish McCallum and Andy Dobson, "Detecting Disease and Parasite Threats to Endangered Species and Ecosystems," *TREE* (Trends in Ecology and Evolution), May 1995; acacia from Andy Dobson and Mick Crawley, "Pathogens and the Structure of Plant Communities," *TREE* (Trends in Ecology and Evolution), October 1994.

17. Donald G. McNeil, Jr., "Predators Get More Than They Bargained For: Tuberculosis," *New York Times,* 15 May 1997; E.D. Anderson, "Morbillivirus Infections in Wildlife (in

Relation to Their Population Biology and Disease Control in Domestic Animals), *Veterinary Microbiology*, vol. 44 (1995), pp. 319–32; Prins and van der Jeugd, op. cit. note 15.

18. Mary Meagher and Margaret E. Meyer, "On the Origin of Brucellosis in Bison of Yellowstone National Park: A Review," *Conservation Biology*, September 1994; for the recent culls and the brucellosis controversy, see Doug Peacock, "The Yellowstone Massacre," *Audubon*, May–June 1997, and Dana Hull, "When the Buffalo Roam They May Not Get Home," *Washington Post*, 22 July 1997; South America from M. Patricia Marchak, *Logging the Globe* (Montreal: McGill-Queen's University Press, 1995), and from Andrew B. Taber, "The Status and Conservation of the Chacoan Peccary in Paraguay," *Oryx*, July 1991; Heather Gardner et al., "Poultry Virus Infection in Antarctic Penguins," *Nature*, 15 May 1997.

19. Douglas H. Chadwick, "The American Prairie: Roots of the Sky," *National Geographic*, October 1993.

20. Introduction of plague and its current status from Edward Tenner, *Why Things Bite Back: Technology and the Revenge of Unintended Consequences* (New York: Alfred A. Knopf, 1996), from Dean Biggins and Jerry Godbey, "Black-Footed Ferrets," in Edward T. LaRoe et al., eds., *Our Living Resources: A Report to the Nation on the Distribution, Abundance, and Health of U.S. Plants, Animals, and Ecosystems* (Washington, DC: U.S. Department of the Interior, National Biological Service, 1995), and from F.M.D. Gulland, "The Impact of Infectious Diseases on Wild Animal Populations—A Review," in B.T. Grenfell and A.P. Dobson, eds., *Ecology of Infectious Diseases in Natural Populations* (Cambridge, U.K.: Cambridge University Press, 1995); prairie dog declines and current control programs from Biggins and Godbey, op. cit. this note, and from Reed Noss, "Cows and Conservation Biology," *Conservation Biology*, September 1994; Brian Miller, Gerardo Ceballos, and Richard Reading, "The Prairie Dog and Biotic Diversity," *Conservation Biology*, September 1994; contemporary plague deaths from "Human Plague," *Washington Post*, 11 July 1997; general danger of diseases to wildlife from Javier A. Simonetti, "Wildlife Conservation Outside Parks Is a Disease-Mediated Task," *Conservation Biology*, April 1995.

21. Ke Chung Kim, "Insect Pests and Evolution," in Kim and McPheron, op. cit. note 7; David Pimentel, "Habitat Factors in New Pest Invasions," in ibid.; rainforest pathogens in tropical agriculture from P.K. Anderson and F.J. Morales, "The Emergence of New Plant Diseases: The Case of Insect-Transmitted Plant Viruses," *Annals of the New York Academy of*

Sciences, Proceedings of a conference held in Woods Hole, MA, on Disease in Evolution: Global Changes and the Emergence of Infectious Diseases, 7–10 November 1993 (New York: 1994). See also examples at the end of Chapter 3.

22. R.C. Shattock, "The Dynamics of Plant Diseases," in J.M. Cherrett and G.R. Sagar, eds., *Origins of Pest, Parasite, Disease and Weed Problems* (Oxford: Blackwell Scientific Publications, 1977); import of the potato, the famine, and suppression of the blight from Jonathan D. Sauer, *Historical Geography of Crop Plants: A Select Roster* (Boca Raton, FL: CRC Press, 1993), and from George N. Agrios, *Plant Pathology,* 4th ed. (San Diego, CA: Academic Press, 1997); recent evolution of the blight from Pat Mooney, "The Hidden 'Hot Zone'—An Epidemic in Two Parts," *Global Pesticide Campaigner* (Pesticide Action Network, North America), December 1995; Anna Maria Gillis, "The Magnificent Devastator Gets Around," *BioScience,* June 1993; William E. Fry and Stephen B. Goodwin, "Re-Emergence of Potato and Tomato Late Blight in the United States," *Plant Disease,* December 1997; Fred Pearce, "The Famine Fungus," *New Scientist,* 26 April 1997; current losses to the blight from Bruce Dorminey, "Suicidal Spuds," *Financial Times,* 22 April 1997.

23. Weeds that became crops from Daniel Zohary and Maria Hopf, *Domestication of Plants in the Old World,* 2nd ed. (Oxford: Clarendon Press, 1994); Bolivian weed potato from OTA, op. cit. note 14; oat rust from Graeme O'Neill, "Duelling Genes: Learning the Game of Disease Resistance," *Ecos,* summer 1996/97.

24. Gary C. Jahn and John W. Beardsley, "Big-Headed Ants, *Pheidole Megacephala:* Interference with the Biological Control of Gray Pineapple Mealy Bugs," in David F. Williams, ed., *Exotic Ants: Biology, Impact, and Control of Introduced Species* (Boulder, CO: Westview Press, 1994).

25. Gregory S. Gilbert and Stephen P. Hubbell, "Plant Diseases and the Conservation of Tropical Forests," *BioScience,* February 1996.

26. Honeybee history from Crosby, op. cit. note 9; current status from Stephen L. Buchman and Gary Paul Nabhan, *The Forgotten Pollinators* (Washington, DC: Island Press/Shearwater Books, 1996).

27. Gary Mead, "Virus Endangers UK Honey Bees," *Financial Times,* 5 August 1997; Buchman and Nabhan, op. cit. note 26; domestic losses from Tina Kelley, "Silence of the Hive Is the Empty Echo of Progress," *Christian Science Monitor,* 9 September 1997, and from Gordon Allen-Wardell et al., "The

Potential Consequences of Pollinator Declines on the Conservation of Biodiversity and Stability of Food Crops Yields," *Conservation Biology,* February 1998.

28. Thomas E. Rinderer, Banjamin P. Oldroyd, and Walter S. Sheppard, "Africanized Bees in the U.S.," *Scientific American,* December 1993.

29. History of the whitefly up until the California infestation from Robert Reinhold, "Potent New Pest Harming Crops in California," *New York Times,* 10 November 1991; spread of the B-biotype from Ke Chung Kim and Bruce A. McPheron, "Biology of Variation: An Epilogue," in Kim and McPheron, op. cit. note 7; current status from Andy Coghlan, "Andes Flower Is Champion Pestkiller," *New Scientist,* 20 January 1996; South America from Lori Ann Thrupp, *Bittersweet Harvests for Global Supermarkets: Challenges in Latin America's Agricultural Export Boom* (Washington, DC: World Resources Institute, 1995); Jane E. Polston and Pamela K. Anderson, "The Emergence of Whitefly-Transmitted Geminiviruses in Tomato in the Western Hemisphere," *Plant Disease,* December 1997; description as most serious pest from Alison Maitland, "Chilean Flower May Join Pest War," *Financial Times,* 31 January 1996.

30. Ryegrass in Australia and occurrences of herbicide-resistant ryegrass from Homer M. LeBaron, "Herbicide Resistance in Plants," in June Fessenden MacDonald, ed., *Biotechnology and Sustainable Agriculture: Policy Alternatives,* National Agricultural Biotechnology Council report 1 (Ithaca, NY: Boyce Thompson Institute for Plant Research, 1989); Roger Cousens and Martin Mortimer, *Dynamics of Weed Populations* (Cambridge, U.K.: Cambridge University Press, 1995); J. Pratley et al., "Glyphosate Resistance in Annual Ryegrass," *Proceedings of the 11th Conference, Grasslands Society of New South Wales,* 1996; other occurrences of herbicide-resistant ryegrass from Ian M. Heap, "The Occurrence of Herbicide-Resistant Weeds Worldwide," *1997 International Survey of Herbicide-Resistant Weeds* (Corvallis, OR: WeedSmart).

31. "Ecological narcotics" from Paul DeBach and David Rosen, *Biological Control by Natural Enemies,* 2nd ed. (Cambridge, U.K.: Cambridge University Press, 1991); 1996 pesticide sales value from "World Agchem Market Recovery Continues," *Agrow,* 11 July 1997; failure to eliminate pests from Kim, op. cit. note 21, and from inference from the 36-year experiment with the herbicide 2,4-D in Canada, described in Roger Cousens and Martin Mortimer, *Dynamics of Weed Populations* (Cambridge, U.K.: Cambridge University Press, 1995); Figure

2–2 from Jodie S. Holt and George P. Georghiou, University of California at Riverside, letter to Peter Weber, Worldwatch Institute, 9 March 1992, with an update on insects from Mark E. Whalon, Department of Entomology, Michigan State University, Ann Arbor, letter to Peter Weber, Worldwatch Institute, 14 February 1994; total number of resistant pests from David Pimentel, professor of insect ecology and agricultural sciences, Cornell University, "Pest Management, Food Security, and the Environment," unpublished paper (Ithaca, NY: Cornell University, March 1995).

32. Cosmopolitan arthropods from Kim, op. cit. note 21; cosmopolitan grain pathogens from David Pimentel, "Habitat Factors in New Pest Invasions," in Kim and McPheron, op. cit. note 7; viral outbreaks from Anderson and Morales, op. cit. note 21; medieval harvest losses from Tenner, op. cit. note 20; current losses from Pimentel, op. cit. note 31, and from George N. Agrios, *Plant Pathology*, 4th ed. (San Diego, CA: Academic Press, 1997) (the 42 percent figure is for major crops); yield increases from David Pimentel et al., "Economic and Environmental Benefits of Biodiversity," *BioScience*, December 1997.

33. Agrios, op. cit. note 32.

34. Total number of agricultural pests from Pimentel et al., op. cit. note 32; corn disease complex from Anderson and Morales, op. cit. note 21.

35. Brian Boag, "Colonials' Revenge: The NZ Flatworm," *Aliens* (newsletter of the Invasive Species Specialist Group, Species Survival Commission, World Conservation Union–IUCN), March 1995; Rowan Taylor et al., *The State of New Zealand's Environment 1997* (Wellington: New Zealand Ministry for the Environment, 1997).

36. Nontraditional export crops from Thrupp, op. cit. note 29; bean virus from Anderson and Morales, op. cit. note 21.

37. National Research Council, Board on Agriculture, *Managing Global Genetic Resources: Agricultural Crop Issues and Policies* (Washington, DC: National Academy Press, 1993).

38. Andean potato crops from Robert J. Frederick, Ivar Virgin, and Eduardo Lindarte, eds., *Environmental Concerns with Transgenic Plants in Centers of Diversity: Potato as a Model* (Stockholm: Stockholm Environment Institute, 1995); karnal bunt from Vandana Shiva, "Species Invasions and the Displacement of Cultural and Biological Diversity," in Odd Terje Sandlund, Peter Johan Schei, and Åslaug Viken, eds., *Proceedings of the Norway/UN Conference on Alien Species, Trondheim, 1–5 July 1996* (Trondheim: Directorate for Nature Management and

Norwegian Institute for Nature Research, 1966), from Morris R. Bonde et al., "Karnal Bunt of Wheat," *Plant Disease,* December 1997, and from the USDA Animal and Plant Health Inspection Service Web site, <http://www.aphis.usda.gov/oa/bunt/kbhome.html>.

39. Hope Shand, "Bio-Meltdown," *New Internationalist,* March 1997.

40. Sandra Blakeslee, "Genetic Engineering Creates Rice Resistant to Destructive Blight," *New York Times,* 15 December 1995; leaf blight from Agrios, op. cit. note 32.

41. A survey of the commercial status of transgenic crops in the United States is available in "What's Coming to Market?" *Gene Exchange* (agricultural biotechnology newsletter of the Union of Concerned Scientists), December 1996; escaped genes from Thomas R. Mikkelsen, Bente Andersen, and Rikke Bagger Jørgensen, "The Risk of Crop Transgene Spread," *Nature,* 7 March 1996, from "Research Confirms Risk of Transgenic Crops," *Gene Exchange* (agricultural biotechnology newsletter of the Union of Concerned Scientists), June 1996, and from Warren E. Leary, "Gene Inserted in Crop Plant Is Shown to Spread to Wild," *New York Times,* 7 March 1996.

42. "USDA Holds National Forum on Bt Resistance," *Gene Exchange* (agricultural biotechnology newsletter of the Union of Concerned Scientists), June 1996; C. Mlot, "Pests Find New Ways around Natural Toxins," *Science News,* 29 November 1997; O. Sarnthoy et al., "Cross-Resistance of *Bacillus thuringiensis* Resistant Population of Diamondback Moth *Plutella xylostella* (Lepidoptera: Yponomeutidae)," *Resistant Pest Management,* winter 1997.

43. Shiva, op. cit. note 38; Richard L. Hill, "Genetic Tinkering with Nature Has Risks," *The Oregonian* (Portland), 8 August 1994.

44. I owe this "revenge" concept to Tenner, op. cit. note 20, who develops it at length in both an ecological and a technical context.

CHAPTER 3. The Forests

1. "Ethiopia, History of" and "Addis Ababa," in *Encyclopaedia Britannica* (15th ed., 1976); teff from Charles F. Rey, *Unconquered Abyssinia as It Is Today* (Philadelphia: J.B. Lippincott, 1924); founding of Addis in 1890 from Ronald J. Horvath, "Addis Ababa's Eucalyptus Forest," *Journal of Ethiopian Studies,* January 1968; Robert Fyfe Zacharin, *Emigrant Eucalypts: Gum Trees as Exotics* (Cape Schanck,

Australia: Khumbila, 1978).

2. Zacharin, op. cit. note 1; "Addis Ababa," op. cit. note 1.
3. Zacharin, op. cit. note 1; for the importance of the Ethiopian Orthodox Church, see "Ethiopia, History of," op. cit. note 1, and "Ethiopian and Somalian Cultures," *Encyclopaedia Britannica* (15th ed., 1976); Menelik's decline from Harold G. Marcus, *Haile Sellassie I: The Formative Years, 1892–1936* (Berkeley: University of California Press, 1987; reprint, Lawrenceville, NJ: Red Sea Press, 1996).
4. "Ethiopia, History of," op. cit. note 1; Marcus, op. cit. note 3; Rey, op. cit. note 1.
5. Zacharin, op. cit. note 1; Horvath, op. cit. note 1, citing P. Mérab, *Impressions d'Ethiopie* (Paris 1921–1929 [sic]), vol. 2; International Conference and Programme for Plant Genetic Resources (ICPPGR), "Country Report for Ethiopia," 1997, at <http://web.icppgr.fao.org/CR/CR/HTML>.
6. Total plantation area from Ashley T. Mattoon, "Paper Forests," *World Watch*, March/April 1998.
7. Figure of 100 million hectares, "or around 5 percent of exploitable forest," from Nigel Dudley, Sue Stolton, and Jean-Paul Jenrenaud, *Pulp Fact: The Environmental and Social Impacts of the Pulp and Paper Industry* (Gland, Switzerland: World Wide Fund for Nature, 1995); U.N. Food and Agriculture Organization (FAO), *State of the World's Forests 1997* (Rome: 1997), gives "net plantation area"—that is, areas planted adjusted by a survival coefficient—in developing countries in 1995 as 81 million hectares, or 4.1 percent of total forested area (plantation area here does not include agricultural plantations, such as rubber, coconut, and oil palm); five- or sixfold increase from Ricardo Carrere and Larry Lohmann, *Pulping the South: Industrial Tree Plantations and the World Paper Economy* (London: Zed Books, 1996); planned doubling from FAO, op. cit. this note.
8. Extent of eucalypt planting from FAO, op. cit. note 7; other plantation species from Theodore Panayotou and Peter S. Ashton, *Not By Timber Alone: Economics and Ecology for Sustaining Tropical Forests* (Washington, DC: Island Press, 1992), and from Julian Evans, *Plantation Forestry in the Tropics* (New York: Oxford University Press, 1992); warm temperate plantations from M. Patricia Marchak, *Logging the Globe* (Montreal: McGill-Queen's University Press, 1995), and, for South Africa, from Carrere and Lohmann, op. cit. note 7.
9. FAO, op. cit. note 7.
10. Extent of lumber production based on Julian Evans, *Plantation Forestry in the Tropics* (New York: Oxford University Press,

1992), and on Panayotou and Ashton, op. cit. note 8; Brazilian steel mills from "Judge Rules Steel Firms in Minas Gerais Must Use Their Own Eucalyptus Trees for Fuel," *International Environment Reporter,* 26 July 1995, and from Econet conference Preservation of Natural Diversity, "Eucalyptus Threatens Biodiversity," posted 24 January 1994; fuelwood production from Panayotou and Ashton, op. cit. note 8, and from Marchak, op. cit. note 8; growth of fiber sector from Carrere and Lohmann, op. cit. note 7; 1995 sales (in excess of value cited here) from Mattoon, op. cit. note 6.

11. Importance of developing-country pulp from Carrere and Lohmann, op. cit. note 7; export trade value from FAO, op. cit. note 7; industrial-country consumption from Mattoon, op. cit. note 6; World Bank funding from Carrere and Lohmann, op. cit. note 7.

12. Carrere and Lohmann, op. cit. note 7; Marchak, op. cit. note 8.

13. Marchak, op. cit. note 8; gulleys from Chris R. Lang, "Problems in the Making: A Critique of Vietnam's Tropical Forestry Action Plan," in Michael J.G. Parnwell and Raymond L. Bryant, eds., *Environmental Change in South-East Asia: People, Politics and Sustainable Development* (London: Routledge, 1996).

14. Carrere and Lohmann, op. cit. note 7.

15. *Maesopsis eminii* from Quentin C.B. Cronk and Janice L. Fuller, *Plant Invaders: The Threat to Natural Ecosystems,* WWF and UNESCO "People and Plants" Conservation Manual 2 (London: Chapman and Hall, 1995); pines and eucalypts from Dave M. Richardson, "Forestry Trees as Alien Invaders: The Current Situation and Prospects for the Future," in Odd Terje Sandlund, Peter Johan Schei, and Åslaug Viken, eds., *Proceedings of the Norway/UN Conference on Alien Species* (Trondheim: Directorate for Nature Management and Norwegian Institute for Nature Research, 1996); eucalypts in the Mediterranean from Vernon H. Heywood, "Patterns, Extents and Modes of Invasions by Terrestrial Plants," in J.A. Drake et al., eds., *Biological Invasions: A Global Perspective,* SCOPE 37 (Chichester, U.K.: Wiley, 1989).

16. C.E. Hughes, "Protocols for Plant Introductions with Particular Reference to Forestry: Changing Perspectives on Risks to Biodiversity and Economic Development," in Charles H. Stirton, chair, *Weeds in a Changing World,* Symposium Proceedings 64 (Farnham, U.K.: British Crop Protection Council, 1996); Stephen J. Pyne, *World Fire: The Culture of Fire on Earth* (New York: Henry Holt, 1995).

17. Frequency of the claim, and the fact that it is false, from

Marchak, op. cit. note 8, from Panayotou and Ashton, op. cit. note 8, and from Carrere and Lohmann, op. cit. note 7; connection between logging and plantations from Dudley, Stolton, and Jenrenaud, op. cit. note 7; Thailand from Larry Lohmann, "Freedom to Plant: Indonesia and Thailand in a Globalizing Pulp and Paper Industry," in Parnwell and Bryant, op. cit. note 13.

18. Perawang from Carrere and Lohmann, op. cit. note 7; Indonesian plantations in general from Dudley, Stolton, and Jeanrenaud, op. cit. note 7.

19. Lohmann, op. cit. note 17; Carrere and Lohmann, op. cit. note 7.

20. Pyne, op. cit. note 16; William D. Montalbano, "On Iberian Peninsula, a Protest Grows over Trees," *Los Angeles Times,* 4 December 1990; Tony Smith, "Eucalyptus Trees Stir Iberian Ecology Alarm as They Soak Up Water," *Los Angeles Times,* 19 November 1989.

21. Thailand from Marchak, op. cit. note 8, from Apichai Puntasen, Somboon Siriprachai, and Chaiyuth Punyasavatsut, "The Political Economy of Eucalyptus: Business, Bureaucracy, and the Thai Government," in Michael C. Howard, ed., *Asia's Environmental Crisis* (Boulder, CO: Westview Press, 1993), and from Carrere and Lohmann, op. cit. note 7; Indonesia from ibid.

22. Laos and Myanmar from Carrere and Lohmann, op. cit. note 7, and from Marchak, op. cit. note 8; China from Carrere and Lohmann, op. cit. note 7; Lohmann, op. cit. note 17; Ken Wilcox, *Chile's Native Forests: A Conservation Legacy* (Redway, CA: Ancient Forest International, 1996).

23. Richardson, op. cit. note 15.

24. *Prosopis* workshop held 13–15 March 1996 at the National Academy of Sciences, Washington, DC; invasions from Cronk and Fuller, op. cit. note 15, and from Shahina A. Ghazanfar, "Invasive *Prosopis* in the Sultanate of Oman," *Aliens* (newsletter of the Invasive Species Specialist Group, Species Survival Commission, World Conservation Union–IUCN), March 1996.

25. A. Niang et al., *"Mimosa scabrella:* A Tree for High Places," *Agroforestry Today,* April-June 1994; invasions from Cronk and Fuller, op. cit. note 15, and from Don C. Schmitz et al., "The Ecological Impact of Nonindigenous Plants," in Daniel Simberloff, Don C. Schmitz, and Tom C. Brown, eds., *Strangers in Paradise: Impact and Management of Nonindigenous Species in Florida* (Washington, DC: Island Press, 1997).

26. Vidya Thakus and C.V. Sarswat, *"Eleagnus umbellata* in the Himalayas," *Agroforestry Today,* April-June 1996; invasions from

John M. Randall, "Exotic Weeds in North American and Hawaiian Natural Areas: The Nature Conservancy's Plan of Attack," in Bill N. McKnight, ed., *Biological Pollution: The Control and Impact of Invasive Exotic Species* (Indianapolis, IN: Indiana Academy of Science, 1993), and from Michael A. Dirr, *Manual of Woody Landscape Plants,* 4th ed. (Champaign, IL: Stipes Publishing, 1990).

27. S.A. Gangoo and T.M. Paul, *"Parrotia jacquemontiana* in India," *Agroforestry Today*, April-June 1996; acacia from Hughes, op. cit. note 16.

28. Vandana Shiva, "Species Invasions and the Displacement of Cultural and Biological Diversity," in Sandlund, Schei, and Viken, op. cit. note 15; Richardson, op. cit. note 15.

29. Native range of alang-alang from P.S. Ramakrishnan and Peter M. Vitousek, "Ecosystem-Level Processes and the Consequences of Biological Invasions," in Drake et al., op. cit. note 15; thatch from Janet Durno, "From 'Imperata Grass Forest' to Community Forest: The Case of Pakhsukjai," *Forests, Trees and People*, September 1996.

30. Alang-alang invasions from Colin L. Sage, "The Search for Sustainable Livelihoods in Indonesian Transmigration Settlements," in Parnwell and Bryant, op. cit. note 13; Indonesia from ibid., and from Marchak, op. cit. note 8; Africa from *IITA Research,* March 1994; Asia overall from National Research Council, *Vetiver Grass: A Thin Green Line against Erosion* (Washington, DC: National Academy Press, 1993); African grasses in the neotropics from Jorge Illueca, "Speech for Trondheim Meeting on Invasive Species," in Sandlund, Schei, and Viken, op. cit. note 15, and from John Tuxill, research fellow, Worldwatch Institute, letter to author, April 1998.

31. Edward H. Forbush and Charles H. Fernald, *The Gypsy Moth* (Boston, MA: Wright and Potter, 1896); Faith Thompson Campbell and Scott E. Schlarbaum, *Fading Forests: North American Trees and the Threat of Exotic Pests* (New York: Natural Resources Defense Council, 1994); U.S. Department of Agriculture (USDA), Forest Service, and Animal and Plant Health Inspection Service (APHIS), *Gypsy Moth Management in the United States: A Cooperative Approach*, draft environmental impact statement, summary (Radnor, PA: May 1995).

32. Steve H. Dreistadt and Donald C. Weber, "Gypsy Moth in the Northeast and Great Lakes States," in Donald L. Dahlsten and Richard Garcia, eds., *Eradication of Exotic Pests: Analysis with Case Histories* (New Haven, CT: Yale University Press, 1989); for moth's feeding preferences, see USDA and APHIS, op. cit.

note 31; Campbell and Schlarbaum, op. cit. note 31.

33. "White moth" from Keith Langdon, biogeographer, U.S. National Park Service, Great Smoky Mountains National Park, discussion with author, 3 March 1997; long-horned beetle from Douglas Martin, "New York's Latest Immigrants: Tree-Eating Beetles from Asia," *New York Times*, 13 September 1996, and from Bruce A. Stein and Stephanie R. Flack, *1997 Species Report Card: The State of U.S. Plants and Animals* (Arlington, VA: The Nature Conservancy, 1997); Michael N. Clout, "Biological Conservation and Invasive Species: The New Zealand Experience," in Sandlund, Schei, and Viken, op. cit. note 15.

34 Original ecology of the chestnut from Gordon G. Whitney, *From Coastal Wilderness to Fruited Plain: A History of Environmental Change in Temperate North America 1500–Present* (Cambridge, U.K.: Cambridge University Press, 1994).

35. Basal area figure from Leslie A. Real, "Sustainability and the Ecology of Infectious Disease," *BioScience*, February 1996; keystone value of the chestnut and spread of the blight from Campbell and Schlarbaum, op cit. note 31, and from George N. Agrios, *Plant Pathology*, 4th ed. (San Diego, CA: Academic Press, 1997).

36. Gretna Weste and G.C. Marks, "The Biology of *Phytophthora cinnamomi* in Australian Forests," *Annual Review of Phytopathology*, vol. 25 (1987), pp. 210, 217; South Africa from Gregory S. Gilbert and Stephen P. Hubbell, "Plant Diseases and the Conservation of Tropical Forests," *BioScience*, February 1996; United States from Campbell and Schlarbaum, op. cit. note 31; Claude Delatour, "View of Pathological Problems in Broadleaved Forest Trees in Europe," in S.P. Raychaudhuri and Karl Maramorosch, eds., *Forest Trees and Palms: Diseases and Control* (Lebanon, NH: Science Publishers, 1996).

37. Jay Cammermeyer, "Life's a Beech—and Then You Die," *American Forests*, July/August 1993; William D. Ostrofsky, "Harvesting Practices, Tree Injuries, and the Management of Forest Health in the Northeastern United States," in Raychaudhuri and Maramorosch, op. cit. note 36.

38. Nematode from Agrios, op. cit. note 35, and from Martin R. Speight and David Wainhouse, *Ecology and Management of Forest Insects* (Oxford, U.K.: Clarendon Press, 1989).

39. Rich Patterson, "Butternut Blues," *American Forests*, July/August 1993; Robert Anderson, "Butternut Canker," *Southern Appalachian Biological Control Initiative Workshop Proceedings*, at <http://www.main.nc.us/SERAMBO/Bcontrol>, viewed July 1997; John Kevin Hiers and Jonathan P. Evans,

"Effects of Anthracnose on Dogwood Mortality and Forest Composition of the Cumberland Plateau (U.S.A.)," *Conservation Biology,* December 1997.

40. Campbell and Schlarbaum, op. cit. note 31.

41. Claude Delatour, "View of Pathological Problems in Broadleaved Forest Trees in Europe," in Raychaudhuri and Maramorosch, op. cit. note 36; Agrios, op. cit. note 35.

42. See, for example, I.A.S. Gibson and T. Jones, "Monoculture as the Origin of Major Forest Pests and Diseases," in J.M. Cherrett and G.R. Sagar, eds., *Parasite, Disease and Weed Problems* (Oxford, U.K.: Blackwell Scientific Publications, 1977); quote from P.J. Kanowski, "Plantation Forestry," in Narenda P. Sharma, ed., *Managing the World's Forests: Looking for Balance Between Conservation and Development* (Dubuque, IA: Kendall/Hunt Publishing, 1992).

43. Tian Guozhong and S.P. Raychaudhuri, "Paulownia Witches' Broom Disease in China: Present Status," in Raychaudhuri and Maramorosch, op. cit. note 36.

44. Carrere and Lohmann, op. cit. note 7.

45. Graeme O'Neill, "Duelling Genes: Learning the Game of Disease Resistance," *Ecos,* summer 1996/97.

46. Carrere and Lohmann, op. cit. note 7.

47. I.A.S. Gibson and T. Jones, "Monoculture as the Origin of Major Forest Pests and Diseases," in Cherrett and Sagar, op. cit. note 42; Carrere and Lohmann, op. cit. note 7.

48. Panayoutou and Ashton, op. cit. note 8.

49. Eucalypts from ibid., from Carrere and Lohman, op. cit. note 7, and from Gibson and Jones, op. cit. note 47; pines and acacias from Carrere and Lohman, op. cit. note 7; for fungi, see, for example, R.T. Paterson and L.M. Mwangi, "Honey Fungus in Agroforestry," *Agroforestry Today,* January–March 1996, and Gilbert and Hubbell, op. cit. note 36.

50. Limits on current knowledge from Gilbert and Hubbell, op. cit. note 36; time lag for insect attacks from Gibson and Jones, op. cit. note 47; infestations from Carrere and Lohmann, op. cit. note 7, and from Marchak, op. cit. note 8.

51. For tree breeding, see, for example, Kanowski, op. cit. note 42, and Anne Simon Moffat, "Moving Forest Trees into the Modern Genetics Era," *Science,* 9 February 1996; for a general statement of the plantation predicament, see Marchak, op. cit. note 8.

52. Danger of pathogen movement out of plantations from Gilbert and Hubbell, op. cit. note 36; Indian and African examples from Carrere and Lohmann, op. cit. note 7, and from Dudley, Stolton, and Jeanrenaud, op. cit. note 7; Weste and Marks, op.

cit. note 36.

53. Mattoon, op. cit. note 6.

54. Richard Snailham, "Ethiopia's New Bloom," *South*, May 1997; status of Ethiopia's forests from ICPPGR, op. cit. note 5; streams around Addis from Sue Edwards, Coordinator, Institute for Sustainable Development, Addis Ababa, Ethiopia, e-mail to author, 6 December 1997; Ethiopia's plantation planning from Carrere and Lohmann, op. cit. note 7.

CHAPTER 4. The Waters

1. Tijs Goldschmidt, *Darwin's Dreampond: Drama in Lake Victoria*, translated by Sherry Marx-Macdonald (Cambridge, MA: The MIT Press, 1996).

2. Importance of the lake fisheries from Richard Ogutu-Ohwayo, "Nile Perch in Lake Victoria: Effects on Fish Species Diversity, Ecosystem Functions and Fisheries," in Odd Terje Sandlund, Peter Johan Schei, and Åslaug Viken, eds., *Proceedings of the Norway/UN Conference on Alien Species, Trondheim, 1–5 July 1996* (Trondheim: Directorate for Nature Management and Norwegian Institute for Nature Research, 1996); development of lake from Thomas C. Johnson et al., "Late Pleistocene Desiccation of Lake Victoria and Rapid Evolution of Cichlid Fishes," *Science*, 23 August 1996, from Carol Kaesuk Yoon, "Lake Victoria's Lightning-Fast Origin of Species," *New York Times*, 27 August 1996, and from Les Kaufman and Peter Ochumba, "Evolutionary and Conservation Biology of Cichlid Fishes as Revealed by Faunal Remnants in Northern Lake Victoria," *Conservation Biology*, September 1993; total number of fish species from Ole Seehaousen, Jacques J.M. van Alphen, and Frans Witte, "Cichlid Fish Diversity Threatened by Eutrophication That Curbs Sexual Selection," *Science*, 19 September 1997.

3. Ogutu-Ohwayo, op. cit. note 2; Goldschmidt, op. cit. note 1.

4. Ogutu-Ohwayo, op. cit. note 2; Goldschmidt, op. cit. note 1.

5. Sardine from Goldschmidt, op. cit. note 1; magnitude of the extinction event from Tijs Goldschmidt, Frans Witte, and Jan Wanink, "Cascading Effects of the Introduced Nile Perch on the Detritivorous/Phytoplanktivorous Species in the Sublittoral Areas of Lake Victoria," *Conservation Biology*, September 1993.

6. Goldschmidt, op. cit. note 1.

7. Ibid.; Ogutu-Ohwayo, op. cit. note 2.

8. Seehaousen, van Alphen, and Witte, op. cit. note 2; anoxia from

Goldschmidt, op. cit. note 1, and from Paul Epstein, "Weeds Bring Disease to the East African Waterways," *The Lancet,* 21 February 1998.

9. Goldschmidt, op. cit. note 1; Ogutu-Ohwayo, op. cit. note 2.

10. Arrival of the hyacinth from Goldschmidt, op. cit. note 1, and from Michela Wrong, "Uganda Faces Tireless Enemy," *Financial Times,* 19 March 1996; rate of growth from Don C. Schmitz et al., "The Ecological Impact and Management History of Three Invasive Alien Aquatic Plant Species in Florida," in Bill N. McKnight, ed., *Biological Pollution: The Control and Impact of Invasive Exotic Species* (Indianapolis, IN: Indiana Academy of Science, 1993). (Water hyacinth is apparently the fastest growing vascular macrophyte in terms of dry biomass.)

11. Wrong, op. cit. note 10; Ogutu-Ohwayo, op. cit. note 2; Epstein, op. cit. note 8; Ann M. Simmons, "Strangling Africa's Regal Lake," *Los Angeles Times,* 28 October 1997.

12. U.S. Department of State cables "Monthly Economic Update [on Uganda]" (February 1995), "Control of Water Hyacinth on Lake Victoria" (December 1992), and "[Defense Intelligence Agency (DIA)] Paper on Lake Victoria" (September 1995), released to the author through a Freedom of Information Act Request, 14 February 1997; "Lake Victoria Cleanup to Begin Thanks to World Bank, GEF Funds," *World Bank News,* 1 August 1996.

13. Population of basin from Ogutu-Ohwayo, op. cit. note 2; for likely decline, see, for example, C.O. Rabuor and J.J. Polovina, "An Analysis of the Multigear, Multispecies Fishery in the Kenyan Waters of Lake Victoria," *Naga: The ICLARM Quarterly,* April 1995, and "[DIA] Paper on Lake Victoria," op. cit. note 12; DIA evaluation from ibid.

14. "Lake Victoria Cleanup," op. cit. note 12.

15. "U.S. Firm Suggests Use of Herbicides to Curb Water Hyacinth in African Lake," *International Environment Reporter,* 9 July 1997; "Government Rejects Proposal by U.S. Firm to Treat Hyacinth Infestation with Chemicals," *International Environment Reporter,* 1 October 1997; "Lake Victoria: The Curse of the Water Hyacinth," *The Economist,* 10 January 1998.

16. Ogutu-Ohwayo, op. cit. note 2; Yvonne Baskin, "Losing a Lake," *Discover,* March 1994.

17. Goldschmidt, op. cit. note 1.

18. Great Lakes profile from Janet Raloff, "From Tough Ruffe to Quagga," *Science News,* 25 July 1992; Great Lakes Panel on Aquatic Nuisance Species, "Biological Invasions" pamphlet (August 1996); Mark Gaten, Communications Officer, Great

Lakes Fisheries Commission, discussion with author, April 1998.

19. Gordon G. Whitney, *From Coastal Wilderness to Fruited Plain: A History of Environmental Change in Temperate North America 1500–Present* (Cambridge, U.K.: Cambridge University Press, 1994); subsequent history from W.A. Pearce et al., "Sea Lamprey *(Petromyzon marinus)* in the Lower Great Lakes," *Canadian Journal of Fisheries and Aquatic Sciences,* November 1980; Carlos M. Fetterolf, Jr., "Why a Great Lakes Fishery Commission and Why a Sea Lamprey International Symposium," *Canadian Journal of Fisheries and Aquatic Sciences,* November 1980; Jon R. Luoma, "Biography of a Lake," *Audubon,* September–October 1996.

20. Luoma, op. cit. note 19.

21. Lamprey from G.J. Farmer, "Biology and Physiology of Feeding in Adult Lampreys," from F.W.H. Beamish, "Biology of the North American Anadromous Sea Lamprey, *Petromyzon marinus,*" and from W.A. Pearce et al., "Sea Lamprey *(Petromyzon marinus)* in the Lower Great Lakes," all in *Canadian Journal of Fisheries and Aquatic Sciences,* November 1980; Edward L. Mills et al., *Exotic Species in the Great Lakes: A History of Biotic Crises and Anthropogenic Introductions,* a Great Lakes Fishery Commission (GLFC) Research Completion Report (Ann Arbor, MI: GLFC, August 1991); A.K. Lamsa et al., "Sea Lamprey *(Petromyzon marinus)* Control—Where to from Here? Report of the SLIS Control Theory Task Force," *Canadian Journal of Fisheries and Aquatic Sciences,* November 1980.

22. Disappearance of predators from L. James Lester, "Marine Species Introductions and Native Species Vitality: Genetic Consequences of Marine Introductions," in M. Richard DeVoe, ed., *Introductions and Transfers of Marine Species: Achieving a Balance Between Economic Development and Resource Protection,* proceedings of a conference and workshop October 30-November 2, 1991, Hilton Head, SC (South Carolina Sea Grant Consortium, 1992); resulting ecological chaos from B.R. Smith and J.J. Tibbles, "Sea Lamprey *(Petromyzon marinus)* in Lakes Huron, Michigan, and Superior: History of Invasion and Control, 1936–78," *Canadian Journal of Fisheries and Aquatic Sciences,* November 1980; alewife from Mills et al., op. cit. note 21; alewife's plankton consumption mentioned in David W. Garton et al., "Biology of Recent Invertebrate Invading Species in the Great Lakes: The Spiny Water Flea, *Bythotrephes cederstroemi* and the Zebra Mussel, *Dreissena polymorpha,*" in McKnight, op. cit. note 10, and in Tammy Keniry and J. Ellen

Marsden, "Zebra Mussels in Southwestern Lake Michigan," in Edward T. LaRoe et al., eds., *Our Living Resources: A Report to the Nation on the Distribution, Abundance, and Health of U.S. Plants, Animals, and Ecosystems* (Washington, DC: U.S. Department of the Interior, National Biological Service, 1995); Edward L. Mills et al., "Exotic Species and the Integrity of the Great Lakes," *BioScience*, November 1994.

23. Fetterolf, op. cit. note 19; lamprey budget from Christina Bjergo et al., "Non-native Aquatic Species in the United States and Coastal Waters," in LaRoe et al., op. cit. note 22; lampricides from F.H. Dahl and R.B. McDonald, "Effects of Control of the Sea Lamprey *(Petromyzon marinus)* on Migratory and Resident Fish Populations," *Canadian Journal of Fisheries and Aquatic Sciences*, November 1980, and from GLFC, *Annual Report of the Great Lakes Fishery Commission 1995* (Ann Arbor, MI: 1995).

24. Dahl and McDonald, op. cit. note 23; GLFC, op. cit. note 23.

25. The Hon. Robert W. Davis in "Hearing before the Subcommittee on Oceanography, Great Lakes and the Outer Continental Shelf and the Subcommittee on Fisheries and Wildlife, Conservation and the Environment of the Committee on Merchant Marine and Fisheries, House of Representatives, One Hundred and Second Congress, First Session, on Status of Efforts to Control Sea Lamprey Populations in the Great Lakes," 17 September 1991 (Washington, DC: U.S. Government Printing Office, 1991).

26. Overview of lake invasions from Edward L. Mills, Spencer R. Hall, and Nijole K. Pauliukonis, "Exotic Species in the Laurentian Great Lakes," *Great Lakes Research Review*, February 1998; zebra mussel origin from Michael L. Ludyanskiy, Derek McDonald, and David MacNeill, "Impact of the Zebra Mussel, a Bivalve Invader," *BioScience*, September 1993; mussel release and spread from Garton et al., op. cit. note 22; quagga from Raloff, op. cit. note 18; ultimate spread of mussel from Amy J. Benson and Charles P. Boydstun, "Invasion of the Zebra Mussel in the United States," in LaRoe et al., op. cit. note 22, and from Ludyanskiy, McDonald, and MacNeill, op. cit. this note.

27. Mussel ecology from Ludyanskiy, McDonald, and MacNeill, op. cit. note 26, and from Paul Arthur Berkman et al., "Zebra Mussels Invade Lake Erie Muds," *Nature*, 7 May 1998; threat to native mussels from Janet N. Abramovitz, *Imperiled Waters, Impoverished Future: The Decline of Freshwater Ecosystems*, Worldwatch Paper 128 (Washington, DC: Worldwatch Institute, March 1996), and from Bruce A. Stein

and Stephanie R. Flack, eds., *America's Least Wanted: Alien Species Invasions of U.S. Ecosystems* (Arlington, VA: The Nature Conservancy, 1996).

28. Glenn Zorpette, "Mussel Mayhem, Continued," *Scientific American*, August 1996.

29. Lake St. Clair from ibid.; mussel predators from Luoma, op. cit. note 19, and from Keniry and Marsden, op. cit. note 22; PCB levels in top predators from Luoma, op. cit. note 19, and from Robert J. Hesselberg and John E. Gannon, "Contaminant Trends in Great Lakes Fish," in LaRoe et al., op. cit. note 22.

30. Early introductions from Mills et al., op. cit. note 22; the $60 million figure includes both the United States and Canada, from Bruce Shupp, Chief, Bureau of Fisheries, New York State Department of Environmental Conservation, in "Hearing before the Subcommittee," op. cit. note 25; lake trout from Michael J. Hansen and James W. Peck, "Lake Trout in the Great Lakes," in LaRoe et al., op. cit. note 22; decline of natives from Abramovitz, op. cit. note 27.

31. Status of lake trout from Hansen and Peck, op. cit. note 30, and from C.C. Krueger, M.L. Jones, and W.W. Taylor, "Restoration of Lake Trout in the Great Lakes: Challenges and Strategies for Future Management," *Journal of Great Lakes Research*, vol. 21 [suppl. 1] (1995), pp. 547–58; exotic salmonids from Mills et al., op. cit. note 22; for diseases affecting salmonids in the Great Lakes, see also Mills et al., op. cit. note 21, and Thomas A. Edsall, Edward L. Mills, and Joseph H. Leach, "Exotic Species in the Great Lakes," in LaRoe et al., op. cit. note 22; alewife from Walter R. Courtenay, Jr., "Biological Pollution through Fish Introductions," in McKnight, op. cit. note 10, and from C.C. Krueger, "Predation of Alewives on Lake Trout Fry in Lake Ontario: Role of an Exotic Species in Preventing Restoration of a Native Species," *Journal of Great Lakes Research*, vol. 21 [suppl. 1] (1995), pp. 458-69.

32. Global extent of intentional aquatic introductions from Peter B. Moyle, "Effects of Invading Species on Freshwater and Estuarine Ecosystems," in Sandlund, Schei, and Viken, op. cit. note 2.

33. Stocking from J.P. Volpe and B.W. Glickman, "It May Be the 'Fish of Kings' But Can British Columbia Afford Atlantic Salmon *(Salmo salar)*?" presentation at the Society for Conservation Biology Annual Meeting, 6–9 June 1997, Victoria, BC, and from Robert R. Stickney, "The Importance of Introduced Marine Species to the Development of the Marine Aquaculture Industry in the United States and Puerto Rico," in DeVoe, op. cit. note 22; escapes from ibid., and from

Courtenay, op. cit. note 31; possible establishment of Atlantic salmon and its effects from Volpe and Glickman, op. cit. this note, and from GREENLines (Grassroots Environmental Effectiveness Network, a project of Defenders of Wildlife), 25 June 1997.

34. Ocean ranching of salmon from Moyle, op. cit. note 32; A.H. Arthington, "Impacts of Introduced and Translocated Freshwater Fishes in Australia," in Sena S. de Silva, ed., *Exotic Aquatic Organisms in Asia,* Special Publication 3 (Manila: Asian Fisheries Society, 1989); Lester, op. cit. note 22; Chile from Sarah Provan, "Chile's Islanders Net an Aquatic Earner," *Financial Times,* 28 February 1996, and from Howard LaFranchi, "Chileans Can't See the Native Forests for the Woodchips," *Christian Science Monitor,* 8 January 1997.

35. Pacific Northwest from Reginald R. Reisenbichler, "Genetic Factors Contributing to Declines of Anadromous Salmonids in the Pacific Northwest," in Deanna J. Stouder et al., eds., *Pacific Salmon and Their Ecosystems* (London: Chapman and Hall, 1997), from Kurt L. Fresh, "The Role of Competition and Predation in the Decline of Pacific Salmon and Steelhead," in ibid., and from Tom Kenworthy, "Fish Hatcheries Caught Between the Wisdom and the Politics of Stocking," *Washington Post,* 1 December 1996; Norway from Moyle, op. cit. note 32; see also D. Gausen and V. Moen, "Large-Scale Escapes of Farmed Atlantic Salmon *(Salmo salar)* into Norwegian Rivers Threaten National Populations," *Canadian Journal of Fisheries and Aquatic Sciences,* vol. 48 (1991), pp. 426–28 (cited in *Naga: The ICLARM Quarterly,* January 1995), and Debora MacKenzie, "Can We Make Supersalmon Safe?" *New Scientist,* 27 February 1996; genetically engineered salmon from ibid.

36. Jack Ganzhorn, J.S. Rohovec, and J.L. Fryer, "Dissemination of Microbial Pathogens Through Introductions and Transfers of Finfish," in Aaron Rosenfield and Roger Mann, eds., *Dispersal of Living Organisms into Aquatic Ecosystems* (College Park, MD: Maryland Sea Grant College, 1992); Emma Solomatina, "Lake Baikal in Close-Up," *Science in Russia,* vol. 3 (1995), p. 50; P.J. Ashton, C.C. Appleton, and P.B.N. Jackson, "Ecological Impacts and Economic Consequences of Alien Invasive Organisms in Southern African Aquatic Ecosystems," in I.A.W. Macdonald, F.J. Kruger, and A.A. Ferrar, eds., *The Ecology and Management of Biological Invasions in Southern Africa* (Cape Town, South Africa: Oxford University Press, 1986); Michael L. Samways, "Southern Hemisphere Insects: Their Variety and the Environmental Pressures upon Them," in Richard Harrington and Nigel E. Stork, eds., *Insects in a Changing*

Environment (London: Academic Press, 1995).

37. Spread of furunculosis from I-Chiu Liao and Hsi-Chiang Liu, "Exotic Aquatic Species in Taiwan," in de Silva, op. cit. note 34; Australia from Angela H. Arthington and David R. Blühdorn, "The Effects of Species Interactions Resulting from Aquaculture Operations," in Donald J. Baird et al., eds., *Aquaculture and Water Resources Management* (Oxford: Blackwell Science, 1996).

38. Hybridization with the cutthroat from Gary H. Thorgaard and Standish K. Allen, "Environmental Impacts of Inbred, Hybrid and Polyploid Aquatic Species," in Rosenfield and Mann, op. cit. note 36; freelance introduction and exotic sport fish as a threat from Constance Holden, "Yellowstone's Cutthroats in Peril," *Science,* 22 March 1996, from Tom Kenworth, "Discovery of Alien Trout Leaves Officials at Yellowstone Reeling," *Washington Post,* 2 October 1994, and from Jim Robbins, "Trouble in Fly Fishermen's Paradise," *New York Times,* 23 August 1996.

39. Widespread introduction from Ganzhorn, Rohovec, and Fryer, op. cit. note 36; Australia and New Zealand from Arthington, op. cit. note 34, from Colin R. Townsend, "Invasion Biology and Ecological Impacts of Brown Trout *Salmo trutta* in New Zealand," *Biological Conservation,* vol. 78 (1996), pp. 13–22, and from Rowan Taylor et al., *The State of New Zealand's Environment 1997* (Wellington: New Zealand Ministry for the Environment, 1997); Dick Russell, "Underwater Epidemic," *The Amicus Journal,* spring 1998; Ganzhorn, Rohovec, and Fryer, op. cit. note 36; Moyle, op. cit. note 32; on the general acclaim for brown trout introductions, see Walter R. Courtenay, Jr. and James D. Williams, "Dispersal of Exotic Species from Aquaculture Sources, with Emphasis on Freshwater Fishes," in Rosenfield and Mann, op. cit. note 36; injury to golden trout from ibid.; injury to the frog from Moyle, op. cit. note 32, and from V.T. Vredenburg and K.R. Matthews, "Introduced Trout and Remaining Populations of Mountain Yellow-Legged Frogs in Sequoia and Kings Canyon National Parks: A Case of Co-Existence or Co-Occurrence?" presentation given at the Society for Conservation Biology Annual Meeting, 6–9 June 1997, Victoria, BC.

40. Overview of Mozambique tilapia's status from Sena S. de Silva, "Exotics—A Global Perspective with Special Reference to Finfish Introductions to Asia," in de Silva, op. cit. note 34; Courtenay and Williams, op. cit. note 39; Geoffrey Fryer, "Biological Invasions in the Tropics: Hypothesis Versus Reality," in P.S. Ramakrishnan, ed., *Ecology of Biological*

Invasions in the Tropics, Proceedings of an International Workshop held at Nainital, India (New Delhi: International Scientific Publications, 1989); tilapia's effects in various countries from Arthington, op. cit. note 34, from Hyan P. Dubey and Afroz Ahmad, "Problems for the Conservation of Freshwater Fish Genetic Resources in India, and Some Possible Solutions," *Naga: The ICLARM Quarterly*, July 1995, and from Rogelio O. Juliano, Rafael Guerrero III, and Inocencio Ronquillo, "The Introduction of Exotic Aquatic Species in the Philippines," in de Silva, op. cit. note 34.

41. Current celebrity status from Joseph J. Molnar et al., "A Global Experiment on Tilapia Aquaculture: Impacts of the Pond Dynamics/Aquaculture CRSP in Rwanda, Honduras, the Philippines and Thailand," *Naga: The ICLARM Quarterly*, April 1996, and from de Silva, op. cit. note 40; invasive abilities of tilapia from Courtenay, op. cit. note 31; adaptation to cooler waters from Jack R. Davidson, James A. Brock, and Leonard G.L. Young, "Introduction of Exotic Species for Aquaculture Purposes," in Rosenfield and Mann, op. cit. note 36, and from Robson A. Collins, "California's Approach to Risk Reduction in the Introduction of Exotic Species," in ibid.; adaptation to seawater from Lester, op. cit. note 22.

42. Asian catfish from Juliano, Guerrero, and Ronquillo, op. cit. note 40; North American bass from Kenji Chiba et al., "Present Status of Aquatic Organisms Introduced into Japan," in de Silva, op. cit. note 34, from Lester, op. cit. note 41, and from Ashton, Appleton, and Jackson, op. cit. note 36; Mekong fish from "Exotic Fish Threaten Local Species," *Asia-Pacific Trade, Environment, and Development Monitor*, 10 October 1997, reprint from *South China Morning Post*, 1 October 1997; pike from A.J. Crivelli, "Are Fish Introductions a Threat to Endemic Freshwater Fishes in the Northern Mediterranean Region?" *Biological Conservation*, vol. 72 (1995), pp. 311–19; Arctic char from Lester, op. cit. note 22; Asian tapeworm from Ganzhorn, Rohovec, and Fryer, op. cit. note 36; carp parasites from Arthington and Blühdorn, op. cit. note 37.

43. Mollusk diseases from C. Austin Farley, "Mass Mortalities and Infectious Lethal Diseases in Bivalve Molluscs and Associations with Geographic Transfers of Populations," in Rosenfield and Mann, op. cit. note 36, and from Arthington and Blühdorn, op. cit. note 37; snail invasions from Rosamond Naylor, "Invasions in Agriculture: Assessing the Cost of the Golden Apple Snail in Asia," *Ambio*, November 1996, and from Peter M. Vitousek et al., "Biological Invasions as Global Environmental Change," *American Scientist*, September/October 1996; bullfrog invasions

from Tan Yo-Jun and Tong He-Yi, "The Status of the Exotic Aquatic Organisms in China," in de Silva, op. cit. note 34, from Liao and Liu, op. cit. note 37, from Twesukdi Piykarnchana, "Exotic Aquatic Species in Thailand," in de Silva, op. cit. note 34, from Begoña Arano, "Species Translocation Menaces Iberian Waterfrogs," *Conservation Biology*, February 1995, and from John Baker, "Gourmet Invader," *Aliens* (Invasive Species Specialist Group, Species Survival Commission, World Conservation Union–IUCN), March 1995.

44. Sport introductions from Moyle, op. cit. note 32, from Robbins, op. cit. note 38, from Nick C. Parker, "Economic Pressures Driving Genetic Changes in Fish," in Rosenfield and Mann, op. cit. note 36, and from Courtenay, op. cit. note 31; carp and pesticides from Edward Tenner, *Why Things Bite Back: Technology and the Revenge of Unintended Consequences* (New York: Alfred A. Knopf, 1996); reservoirs and ubiquity of exotics from Moyle, op. cit. note 32; tailwaters from Courtenay, op. cit. note 31.

45. Figure 4–2 from Anne Platt McGinn, *Rocking the Boat: Conserving Fisheries and Protecting Jobs*, Worldwatch Paper 142 (Washington, DC: Worldwatch Institute, June 1998); U.S. Bureau of the Census, *International Data Base*, electronic database, Suitland, MD, updated 10 October 1997.

CHAPTER 5. Islands

1. John L. Culliney, *Islands in a Far Sea: Nature and Man in Hawaii* (San Francisco, CA: Sierra Club Books, 1988).

2. Tourists from Alan Holt, "An Alliance of Biodiversity, Agriculture, Health, and Business Interests for Improved Alien Species Management in Hawaii," in Odd Terje Sandlund, Peter Johan Schei, and Åslaug Viken, eds., *Proceedings of the Norway/UN Conference on Alien Species, Trondheim, 1–5 July 1996* (Trondheim: Directorate for Nature Management and Norwegian Institute for Nature Research, 1996).

3. An overview of the Hawaiian biota is available in Lucius G. Eldredge and Scott E. Miller, "How Many Species Are There in Hawaii?" *Bishop Museum Occasional Paper 41* (Honolulu, HI: 1995) and supplement in *Occasional Paper 45* (Honolulu, HI: 1996), and in Bishop Museum Hawaii Biological Survey (HBS), "Hawaii's Endangered Species," at <http://www.bishop.hawaii.org/bishop/HBS/>, updated August 1997.

4. Lobelias and honeycreepers from Elizabeth Royte, "On the

Brink: Hawaii's Vanishing Species," *National Geographic*, September 1995, and from Culliney, op. cit. note 1; M.G. Hadfield and L.J. Hadway, "Invasions and Extinctions: Case Histories from Hawaiian Tree Snails," paper presented at the annual meeting of the Society for Conservation Biology (Victoria, BC: June 1997). Violets and ferns from Culliney, op. cit. note 1.

5. HBS, op. cit. note 3, gives 8,850 as the number of endemic species; 10,000 figure from Alan Holt, "An Alliance of Biodiversity, Agriculture, Health, and Business Interests for Improved Alien Species Management in Hawaii," in Sandlund, Schei, and Viken, op. cit. note 2; Charles P. Stone and Lloyd L. Loope, "Alien Species in Hawaiian National Parks," in W.L. Halvorson and G.E. Davis, eds., *Science and Ecosystem Management in the National Parks* (Tucson: University of Arizona Press, 1996).

6. HBS, op. cit. note 3; Holt, op. cit. note 5.

7. General point regarding logging from Culliney, op. cit. note 1; eucalyptus and melaleuca from Peter Ward, *The End of Evolution: On Mass Extinctions and the Preservation of Biodiversity* (New York: Bantam Books, 1994); general importance of exotic predators from Holt, op. cit. note 5, from Culliney, op. cit. note 1, and from Christopher Lever, *Naturalized Mammals of the World* (London: Longman, 1985).

8. Ward, op. cit. note 7; Culliney, op. cit. note 1; C.T. Atkinson et al., "Wildlife Disease and Conservation in Hawaii: Pathogenicity of Avian Malaria *(Plasmodium relictum)* in Experimentally Infected Iiwi *(Vestiaria coccinea)*," *Parasitology*, vol. 111 (1995), pp. s59–s69; birds dropping out of trees from Royte, op. cit. note 4; plants and seabird habitat from Culliney, op. cit. note 1; displacement from Gordon H. Orians, "Thought for the Morrow: Cumulative Threats to the Environment," *Environment*, September 1995; exotic bird numbers from HBS, op. cit. note 3.

9. Bruce A. Stein and Stephanie R. Flack, eds., *America's Least Wanted: Alien Species Invasions of U.S. Ecosystems* (Arlington, VA: The Nature Conservancy, 1996).

10. Pigs and goats from Royte, op. cit. note 4; goats and game birds from Culliney, op. cit. note 1; grasses from Quentin C.B. Cronk and Janice L. Fuller, *Plant Invaders: The Threat to Natural Ecosystems*, WWF and UNESCO "People and Plants" Conservation Manual 2 (London: Chapman and Hall, 1995), and from Carla M. D'Antonio and Peter M. Vitousek, "Biological Invasions by Exotic Grasses, the Grass/Fire Cycle, and Global Change," *Annual Review of Ecology and Systematics*,

vol. 23 (1992), pp. 63-87; lack of pollination and exotic insects from Ward, op. cit. note 7; exotic insects also from Francis G. Howarth, Gordon Nishida, and Adam Asquith, "Insects of Hawaii," in Edward T. LaRoe et al., eds., *Our Living Resources: A Report to the Nation on the Distribution, Abundance, and Health of U.S. Plants, Animals, and Ecosystems* (Washington, DC: U.S. Department of the Interior, National Biological Service, 1995), and from Neil J. Reimer, "Distribution and Impact of Alien Ants in Vulnerable Hawaiian Ecosystems," in David F. Williams, ed., *Exotic Ants: Biology, Impact, and Control of Introduced Species* (Boulder, CO: Westview Press, 1994); long-legged ant from ibid.

11. Koster's curse from Ward, op. cit. note 7, and from Cronk and Fuller, op. cit. note 10; fire tree and banana poka from ibid.; number of exotic plants from HBS, op. cit. note 3.

12. Holt, op. cit. note 5.

13. Brown tree snake from Stein and Flack, op. cit. note 9, and from Gordon H. Rodda, Thomas H. Fritts, and David Chiszar, "The Disappearance of Guam's Wildlife," *BioScience,* October 1997; other extinctions, suppressions, and current prey base from T.H. Fritts and G.H. Rodda, "Invasions of the Brown Tree Snake," in LaRoe et al., op. cit. note 10.

14. Possibility of colonization from G. Perry, G. Rodda, and T. Fritts, "Biological, Economic and Political Factors Relevant to Future Spread of Brown Tree Snakes *(Boiga irregularis)* in the Pacific," paper presented at the annual meeting of the Society for Conservation Biology (Victoria, BC: June 1997); William Claiborne, "Trouble in Paradise?" *Washington Post,* 23 August 1997.

15. Cronk and Fuller, op. cit. note 10; Stein and Flack, op. cit. note 9; Jean-Yves Meyer, "Status of *Miconia calvescens* (Melastomataceae), a Dominant Invasive Tree in the Society Islands (French Polynesia)," *Pacific Science,* vol. 50, no. 1 (1996).

16. K.V. Sykora, "History of the Impact of Man on the Distribution of Plant Species," in F. di Castri, A.J. Hansen, and M. Debussche, eds., *Biological Invasions in Europe and the Mediterranean Basin* (Dordrecht, the Netherlands: Kluer Academic Publishers, 1990); rate of arrival from Michael N. Clout, "Biological Conservation and Invasive Species: The New Zealand Experience," in Sandlund, Schei, and Viken, op. cit. note 2; clematis from Cronk and Fuller, op. cit. note 10.

17. Native bats from A.F. Mark and G.D. McSweeney, "Patterns of Impoverishment in Natural Communities: Case History Studies in Forest Ecosystems—New Zealand," in George M.

Woodwell, ed., *The Earth in Transition: Patterns and Processes of Biotic Impoverishment* (Cambridge, U.K.: Cambridge University Press, 1990); exotic mammals from Clout, op. cit. note 16; brush-tailed possum from A.B. Rose, C.J. Pekelharing, and K.H. Platt, "Magnitude of Canopy Dieback and Implications for Conservation of Southern Rata-Kamahi *(Metrosideros umbellata – Weinmannia racemosa)* Forests, Central Westland, New Zealand," *New Zealand Journal of Ecology*, vol. 16, no.1 (1992), from Clout, op. cit. note 16, and from Helen Goss, "The Mysterious Case of the Wobbly Possum," *New Scientist*, 5 August 1995.

18. Number of threatened life forms (800 species and 200 sub-species) from Rowan Taylor et al., *The State of New Zealand's Environment 1997* (Wellington: New Zealand Ministry for the Environment, 1997); invasion as the most serious threat from Clout, op. cit. note 16.

19. Loss of native mammals from Craig MacFarland and Miguel Cifuentes, "Case Study: Ecuador," in Victoria Dompka, ed., *Human Population, Biodiversity and Protected Areas: Science and Policy Issues*, Report of a Workshop April 20–21, 1995 (Washington, DC: American Association for the Advancement of Science, 1996); Deborah A. Clark, "Native Land Mammals," in Roger Perry, ed., *Galapagos* (New York: Pergamon Press, 1984); current exotic mammal damage from Ian A.W. Macdonald et al., "Wildlife Conservation and the Invasion of Nature Reserves by Introduced Species: A Global Perspective," in J.A. Drake et al., eds., *Biological Invasions: A Global Perspective*, SCOPE 37 (Chichester, U.K.: John Wiley and Sons, 1989), and from Paul A. Stone, Howard L. Snell, and Heidi M. Snell, "Behavioral Diversity as Biological Diversity: Introduced Cats and Lava Lizard Wariness," *Conservation Biology*, June 1994; ants from Ines de la Vega, "Food Searching Behavior and Competition Between *Wasmannia Auropunctata* and Native Ants on Santa Cruz and Isabela, Galapagos Islands," in Williams, op. cit. note 10.

20. Cronk and Fuller, op. cit. note 10; "Galápagos Fires," *Oryx*, January 1995; *IUCN Bulletin*, no. 3, 1994; A. Mauchamp, "Threat from Alien Plant Species in the Galápagos Islands," *Conservation Biology*, February 1997.

21. Wendy Strahm, "Invasive Species in Mauritius: Examining the Past and Charting the Future," in Sandlund, Schei, and Viken, op. cit. note 2; Cronk and Fuller, op. cit. note 10; Stein and Flack, op. cit. note 9.

22. Charles Darwin, *Journal of Researches into the Geology and Natural History of the Various Countries Visited by H.M.S. Beagle*,

under the Command of Captain Fitzroy, R.N. from 1832 to 1836 (London: 1839).

23. "Naive" island fauna from Hadfield and Hadway, op. cit. note 4; Hawaiian plants from Royte, op. cit. note 4.

24. New Zealand from Clout, op. cit. note 16; Hawaii from Peter M. Vitousek, "Biological Invasions and Ecosystem Properties: Can Species Make a Difference?" in Harold A. Mooney and James A. Drake, eds., *Ecology of Biological Invasions of North America and Hawaii*, Ecological Studies: Analysis and Synthesis 58 (New York: Springer Verlag, 1986).

25. Seabirds from Douglas F. Stotz et al., *Neotropical Birds: Ecology and Conservation* (Chicago: University of Chicago Press, 1996); 57 percent from James H. Brown, "Patterns, Modes and Extents of Invasions by Vertebrates," in Drake et al., op. cit. note 19; 70 percent cited in John Balzar, "A Deadly Plague of Stowaways," *Los Angeles Times*, 17 May 1993.

26. General extinction percentages from Robert L. Peters and Thomas E. Lovejoy, "Terrestrial Fauna," in B.L. Turner II, ed., *The Earth as Transformed by Human Action: Global and Regional Changes in the Biosphere over the Past 300 Years* (Cambridge, U.K.: Cambridge University Press and Clark University, 1990).

27. Stuart L. Pimm et al., "The Future of Biodiversity," *Science*, 21 July 1995.

28. Michael A. Huston, *Biological Diversity: The Coexistence of Species on Changing Landscapes* (Cambridge, U.K.: Cambridge University Press, 1994); Michael L. Rosenzweig, *Species Diversity in Space and Time* (Cambridge, U.K.: Cambridge University Press, 1995).

29. Pimm et al., op. cit. note 27; the most recent authoritative survey of endangered fauna is Jonathan Baillie and Brian Groombridge, *1996 IUCN Red List of Threatened Animals* (Gland, Switzerland: International Union for Conservation of Nature and Natural Resources, 1996); continental extinction count from Andrew P. Dobson, *Conservation and Biodiversity* (New York: Scientific American Library, 1996).

30. Dobson, op. cit. note 29.

31. Geography of the Kingdom and plant flocks from Huston, op. cit. note 28; Rosenzweig, op. cit. note 28; species diversity and endangerment count from Brian J. Huntley, "South Africa's Experience Regarding Alien Species: Impacts and Controls," in Sandlund, Schei, and Viken, op. cit. note 2.

32. Argentine ant from N.E. Stork and M.J. Samways, "Inventorying and Monitoring of Biodiversity," in V.H. Heywood, ed., *Global Biodiversity Assessment* (Cambridge, U.K.: Cambridge University Press and the United Nations

Environment Programme, 1995).

33. For an overview of global conversion rates, see John F. Richards, "Land Transformation," in Turner, op. cit. note 26; halving of forest cover from Dirk Bryant, Daniel Nielsen, and Laura Tangley, *The Last Frontier Forests: Ecosystems and Economies on the Edge* (Washington, DC: World Resources Institute, 1997); current rates of forest loss from ibid. and from U.N. Food and Agriculture Organization, *State of the World's Forests 1997* (Rome: 1997); forest loss in earlier eras is extrapolated from Richards, op. cit. in this note.

34. William D. Newmark, "Extinction of Mammal Populations in Western North American National Parks," *Conservation Biology*, June 1995; Dobson, op. cit. note 29; Michael L. Rosenzweig and Colin W. Clark, "Island Extinction Rates from Regular Censuses," *Conservation Biology*, June 1994.

35. I.M. Turner, "Species Loss in Fragments of Tropical Rain Forest: A Review of the Evidence," *Journal of Applied Ecology*, vol. 33, no. 2 (1996); Newmark, op. cit. note 34.

36. The review was undertaken by the SCOPE Working Group on Nature Reserves. See Michael B. Usher, "Biological Invasions into Tropical Nature Reserves," in P.S. Ramakrishnan, ed., *Ecology of Biological Invasions in the Tropics*, Proceedings of an International Workshop Held at Nainital, India (New Delhi: International Scientific Publications, 1989); U.S. parks from Peter M. Vitousek et al., "Biological Invasions as Global Environmental Change," *American Scientist*, September/ October 1996.

37. Brian Czech and Paul R. Krausman, "Distribution and Causation of Species Endangerment in the United States," *Science*, 22 August 1997; Don C. Schmitz and Daniel Simberloff, "Biological Invasions: A Growing Threat," *Issues in Science and Technology*, summer 1997 (the lower percentage (from Czech and Krausman) may exclude some cases of hybridization with exotics and infection by exotic pathogens); invasion as second most important cause of endangerment from Schmitz and Simberloff, op. cit. this note. Table 5–3 based on Macdonald et al., op. cit. note 19, as adapted in Jeffrey A. McNeely, "The Great Reshuffling: How Alien Species Help Feed the Global Economy," in Sandlund, Schei, and Viken, op. cit. note 2, with freshwater fish from John Tuxill, *Losing Strands in the Web of Life: Vertebrate Declines and the Conservation of Biological Diversity*, Worldwatch Paper 141 (Washington, DC: Worldwatch Institute, May 1998).

38. Figure 5–1 from Peter M. Vitousek et al., "Biological Invasions as Global Environmental Change," *American Scientist*,

September 1996, and (for the United States) from Center for Plant Conservation, "Plants in Peril," <http://www.mobot.org/CPC/welcome.html>; Jane H. Bock and Carl E. Bock, "The Challenges of Grassland Conservation," in Anthony Joern and Kathleen H. Keeler, eds., *The Changing Prairie: North American Grasslands* (New York: Oxford University Press, 1995); Bruce Coblentz, professor of fisheries and wildlife, Oregon State University, Corvallis, OR, e-mail to author, September 1995.

39. Jeff Crooks and Michael E. Soulé, "Lag Times in Population Explosions of Invasive Species: Causes and Implications," in Sandlund, Schei, and Viken, op. cit. note 2.

CHAPTER 6. Colonists

1. Edwin Way Teale, "Bird Invasion," in Paul S. Eriksson and Alan Pistorius, eds., *Treasury of North American Bird Lore* (Middlebury, VT: Paul S. Eriksson, 1987), excerpted from Edwin Way Teale, *Days without Time* (New York: Dodd Mead, 1948); Frank M. Chapman, "The European Starling as an American Citizen," *Natural History*, September/October 1925; Florida and the prairies from Ted Morgan, *Wilderness at Dawn: The Settling of the North American Continent* (New York: Simon and Schuster, 1993), and from Gordon G. Whitney, *From Coastal Wilderness to Fruited Plain: A History of Environmental Change in Temperate North America 1500–Present* (Cambridge, U.K.: Cambridge University Press, 1994); starling spread across North America from Christopher Feare, *The Starling* (New York: Oxford University Press, 1984).

2. Schieffelin's biography from Edward Tenner, *Why Things Bite Back: Technology and the Revenge of Unintended Consequences* (New York: Alfred A. Knopf, 1996); aims of the American Acclimatization Society from Christopher Lever, *They Dined on Eland: The Story of the Acclimatisation Societies* (London: Quiller Press, 1992); starling as Schieffelin's only successful introduction from Chapman, op. cit. note 1.

3. Standard view of Schieffelin from Teale, op. cit. note 1; lack of documentary evidence from Tenner, op. cit. note 2; Friends of Shakespeare from Lever, op. cit. note 2; Paul R. Ehrlich, David S. Dobkin, and Darryl Wheye, *The Birder's Handbook: A Field Guide to the Natural History of North American Birds* (New York: Simon and Schuster, 1988).

4. Bradford cited in Charles E. Little, *The Dying of the Trees: The Pandemic in America's Forests* (New York: Viking, 1995).

5. "Neo-Europe" term from Alfred W. Crosby, *Ecological*

Imperialism: The Biological Expansion of Europe, 900–1900
(Cambridge, U.K.: Cambridge University Press, 1986); Robert
R. Stickney, *Aquaculture in the United States: A Historical Survey*
(New York: John Wiley and Sons, 1996).

6. Stickney, op. cit. note 5; cosmopolitan distribution from Jack
   Ganzhorn, J.S. Rohovec, and J.L. Fryer, "Dissemination of
   Microbial Pathogens Through Introductions and Transfers of
   Finfish," in Aaron Rosenfield and Roger Mann, eds., *Dispersal
   of Living Organisms into Aquatic Ecosystems* (College Park, MD:
   Maryland Sea Grant College, 1992); suppression of native fish
   from Walter R. Courtenay, Jr., and James D. Williams,
   "Dispersal of Exotic Species from Aquaculture Sources, with
   Emphasis on Freshwater Fishes," in ibid., and from H.P.C.
   Shetty, M.C. Nandeesha, and A.G. Jhingran, "Impact of Exotic
   Aquatic Species in Indian Waters," in Sena S. de Silva, ed.,
   Exotic Aquatic Organisms in Asia, Asian Fisheries Society (AFS)
   Special Publication 3 (Manila: AFS, 1989).

7. Stickney, op. cit. note 5.

8. Lever, op. cit. note 2.

9. Ibid.

10. Ibid.

11. Failed introductions from ibid.; rabbits from K. Myers,
    "Introduced Vertebrates in Australia, with Emphasis on the
    Mammals," in R.H. Groves and J.J. Burdon, eds., *Ecology of
    Biological Invasions* (Cambridge, U.K.: Cambridge University
    Press, 1986), from Christopher Lever, *Naturalized Mammals of
    the World* (London: Longman, 1985), and from George
    Laycock, *The Alien Animals: The Story of Imported Wildlife*
    (Garden City, NY: Natural History Press, 1966).

12. Birds from Christopher Lever, *Naturalized Birds of the World*
    (New York: Wiley, 1987), and from Lever, op. cit. note 2; deer
    from Lever, op. cit. note 11; prickly pear from Lever, op. cit.
    note 2.

13. Michael A. Osborne, *Nature, the Exotic, and the Science of French
    Colonialism* (Bloomington: Indiana University Press, 1994).

14. First quote from Lever, op. cit. note 2; second quote from
    Osborne, op. cit. note 13.

15. Zootechnie and Burchell's zebra from Warwick Anderson,
    "Climates of Opinion: Acclimatization in Nineteenth-Century
    France and England," *Victorian Studies*, winter 1992; game
    birds (which were first introduced well before the advent of the
    societies) from Paul Isenmann, "Some Recent Bird Invasions in
    Europe and the Mediterranean Basin," in F. di Castri, A.J.
    Hansen, and M. Debussche, eds., *Biological Invasions in Europe
    and the Mediterranean Basin* (Boston, MA: Kluwer Academic

Publishers, 1990), and from Lever, op. cit. note 2; fish and deer from ibid.; deer also from P.R. Ratcliffe, "The Control of Red and Sika Deer Populations in Commercial Forests," and M.J. Hannan and J. Whelan, "Deer and Habitat Relations in Managed Forests," both in R.J. Putman, ed., *Mammals as Pests* (London: Chapman and Hall, 1989).

16. Botanical garden network from Anderson, op. cit. note 15; pattern of exotic cultivation from J.W. Purseglove, "History and Functions of Botanic Gardens with Special Reference to Singapore," *Garden's Bulletin Singapore*, vol. 17 (1959), pp. 125–54, cited in Vernon H. Heywood, "Patterns, Extents and Modes of Invasions by Terrestrial Plants," in J.A. Drake et al., eds., *Biological Invasions: A Global Perspective*, SCOPE 37 (Chichester, U.K.: Wiley, 1989).

17. Figures 6–1 and 6–2 from Ingo Kowarik, "Time Lags in Biological Invasions with Regard to the Success and Failure of Alien Species," in Petr Pyšek et al., eds., *Plant Invasions: General Aspects and Special Problems* (Amsterdam: Academic Publishing, 1995).

18. Anderson, op. cit. note 15; Lever, op. cit. note 2.

19. Lever, op. cit. note 2; Tenner, op. cit. note 2; Lever, op. cit. note 11.

20. Deer from George A. Feldhamer and William E. Armstrong, "Interspecific Competition Between Four Exotic Species and Native Artiodactyls in the United States," *Transactions of the 58th North American Wildlife and Natural Resources Conference* (1993); Victor B. Scheffer, "The Olympic Goat Controversy: A Perspective," *Conservation Biology*, December 1993; pheasant from Richard E. Warner, "Philosophical and Ecological Perspecites of Highly Valued Exotic Animals: Case Studies of Domestic Cats and Game Birds," and Francis M. Harty, "How Illinois Kicked the Exotic Habit," both in Bill N. McKnight, ed., *Biological Pollution: The Control and Impact of Invasive Exotic Species* (Indianapolis, IN: Indiana Academy of Science, 1993).

21. Mink from John Timson, "Murderous Mink Wipe Out Water Voles," *New Scientist*, 30 March 1991; nutria from Christopher Cooper, "Louisiana Is Trying to Turn Pest Into a Meal," *New York Times*, 14 December 1997, from Bruce Coblentz, professor of fisheries and wildlife, Oregon State University, Corvallis, OR, e-mail to author, April 1998, and from Laycock, op. cit. note 11; muskrat from ibid. and from Mark Williamson, *Biological Invasions* (London: Chapman and Hall, 1996); beaver from Ken Wilcox, *Chile's Native Forests: A Conservation Legacy* (Redway, CA: Ancient Forest International, 1996).

22. Darius is quoted in R.H. Groves, "The Biogeography of

Mediterranean Plant Invasions," in R.H. Groves and F. di
Castri, eds., *Biogeography of Mediterranean Invasions*
(Cambridge, U.K.: Cambridge University Press, 1991); Loeb
Classical Library, Pliny, *Natural History, vol. 4* (Cambridge,
MA: Harvard University Press, 1986); Manuel Komroff, ed.,
The Travels of Marco Polo (New York: Random House, 1926).

23. Table 6–2 based on the following: rubber vine from Stella E.
Humphries, "Invasive Plants in Australia," *Aliens* (Invasive
Species Specialist Group, Species Survival Commission, World
Conservation Union–IUCN), March 1995; clematis from
Quentin C.B. Cronk and Janice L. Fuller, *Plant Invaders: The
Threat to Natural Ecosystems,* WWF and UNESCO "People and
Plants" Conservation Manual 2 (London: Chapman and Hall,
1995); water hyacinth from J.C. Joyce, "Practical Uses of
Aquatic Weeds," in Arnold H. Pieterse and Kevin J. Murphy,
eds., *Aquatic Weeds: The Ecology and Management of Nuisance
Aquatic Vegetation* (Amsterdam: Academic Publishing, 1995),
and from B. Gopal, "Aquatic Weed Problems and Management
in Asia," in ibid.; purple loosestrife from Keith R. Edwards,
Michael S. Adams, and Jan Kvet, "Invasion History and
Ecology of *Lythrum salicaria* in North America," in Pyšek et al.,
op. cit. note 17, and from Sarah Reichard and Faith Campbell,
"Invited but Unwanted," *American Nurseryman,* 1 July 1996;
knotweeds from John P. Bailey, Lois E. Child, and Max Wade,
"Assessment of the Genetic Variation and Spread of British
Populations of *Fallopia japonica* and Its Hybrid *Fallopia x
bohemica*," in Pyšek et al., op. cit. note 17, and from Leslie A.
Seiger, "*Fallopia japonica*: Japanese Knotweed," in John M.
Randall and Janet Marinelli, eds., *Invasive Plants: Weeds of the
Global Garden* (New York: Brooklyn Botanic Garden, 1996);
saltcedars from Peter Friederici, "The Alien Saltcedar,"
American Forests, January/February 1995, and from William
Wiesenborn, "*Tamarix ramosissima, T. chinensis, T. parvifloria*:
Tamarisk," in Randall and Marinelli, op. cit. this note. Survey
from Pierre Binggeli, "A Taxonomic, Biogeographical and
Ecological Overview of Invasive Woody Plants," *Journal of
Vegetation Science*, vol. 7 (1996), p. 124. Natural areas weeds in
North America from Janet Marinelli, "Redefining the Weed," in
Randall and Marinelli, op. cit. this note.

24. Percentage of invasive plants in nursery trade from Faith
Campbell, exotic species expert, Western Ancient Forests
Campaign, letter to author, April 1998; senecio from Peter
Williams, Landcare Research, Nelson, New Zealand, e-mail to
U.S. Department of Agriculture (USDA) Animal and Plant
Health Inspection Service (APHIS) weeds listserver

<weeds@info.aphis.usda.gov>, 2 December 1997, and from Mike Harre, Auckland Regional Council, Auckland, New Zealand, e-mail to USDA APHIS weeds listserver, 3 December 1997.

CHAPTER 7. Accidents

1. Andrew N. Cohen and James T. Carlton, *Nonindigenous Aquatic Species in a United States Estuary: A Case Study of the Biological Invasions of the San Francisco Bay and Delta* (Springfield, VA: National Technical Information Service, 1995).
2. A.N. Cohen, J.T. Carlton, and M.C. Fountain, "Introduction, Dispersal and Potential Impacts of the Green Crab *Carcinus maenas* in San Francisco Bay, California," *Marine Biology*, vol. 122 (1995), pp. 225–26; Andrew N. Cohen, "Have Claw, Will Travel," *Aquatic Nuisance Species Digest*, August 1997; Joel W. Hedgpeth, "Foreign Invaders," *Science*, 2 July 1993; Cohen and Carlton, op. cit. note 1; William K. Stevens, "Bay Inhabitants Are Mostly Aliens," *New York Times*, 20 August 1996.
3. Andrew N. Cohen, "Chinese Mitten Crabs in North America," *Aquatic Nuisance Species Digest*, November 1995; Gordy Slack, "Chinese Crabs Discover San Francisco Bay," *Pacific Discovery*, summer 1996; Cohen and Carlton, op. cit. note 1.
4. Cohen and Carlton, op. cit. note 1; James T. Carlton, "Ballast Water," *Aliens* (Invasive Species Specialist Group, Species Survival Commission, World Conservation Union–IUCN), March 1995; Peter Tirschwell, "Stemming the Tide of Change," *Journal of Commerce*, 24 June 1996; "Western [sic] Clam Poses Threat, Study Finds," *New York Times*, 16 March 1997.
5. Cohen and Carlton, op. cit. note 1.
6. Committee on Ships' Ballast Operations, National Research Council, *Stemming the Tide: Controlling Introductions of Nonindigenous Species by Ships' Ballast Water* (Washington, DC: National Academy Press, 1996).
7. Swimming pool ratio derived from Christopher J. Wiley, "Aquatic Nuisance Species: Nature, Transport, and Regulation," in Frank D'Itri, ed., *Zebra Mussels and Other Aquatic Nuisance Species* (Chelsea, MI: Ann Arbor Press, 1997).
8. Committee on Ships' Ballast Operations, op. cit. note 6.
9. Table 7–1 based on the following: outbound from Europe from Chapter 4, from U.S. Congress, Office of Technology Assessment (OTA), *Harmful Nonindigenous Species in the United States* (Washington, DC: September 1993), and from Chris

Viney, "Pest Control in the Deep," *Ecos,* spring 1996; outbound from the Americas from this Chapter, from Committee on Ships' Ballast Operations, op. cit. note 6, from James T. Carlton and Jonathan B. Geller, "Ecological Roulette: The Global Transport of Nonindigenous Marine Organisms, *Science,* 2 July 1993, and from Jeffrey A. McNeely, "The Great Reshuffling: How Alien Species Help Feed the Global Economy," in Odd Terje Sandlund, Peter Johan Schei, and Åslaug Viken, eds., *Proceedings of the Norway/UN Conference on Alien Species, Trondheim, 1–5 July 1996* (Trondheim: Directorate for Nature Management and Norwegian Institute for Nature Research, 1996); outbound from East Asia from this Chapter, from Carlton and Geller, op. cit. this note, from Ian Anderson, "Stowaway Drives Fish to Brink of Extinction," *Aliens* (Invasive Species Specialist Group, Species Survival Commission, IUCN), September 1995, from David Malakoff, "Extinction on the High Seas," *Science,* 25 July 1997, from Viney, op. cit. this note, from James T. Carlton, "Marine Bioinvasions: The Alteration of Marine Ecosystems by Nonindigenous Species," *Oceanography,* vol. 9, no.1 (1996), and from J.C. Sanderson, "A Preliminary Survey of the Distribution of the Introduced Macroalga, *Undaria pinnatifida* (Harvey) Suringer on the East Coast of Tasmania, Australia," *Botanica Marina,* vol. 33 (1990); outbound from South Asia from James T. Carlton, "Man's Role in Changing the Face of the Ocean: Biological Invasions and Implications for Conservation of Near-Shore Environments," *Conservation Biology,* September 1989, from Carlton and Geller, op. cit. this note, and from James T. Carlton, "Four Species of Marine Crabs Invade North America," *Aliens* (Invasive Species Specialist Group, Species Survival Commission, IUCN), September 1995; outbound from Australia from Tim Allen, "Ballast Water Blues," *Habitat Australia,* November 1994.

10. Ralph T. Bryan, "Alien Species and Emerging Infectious Diseases: Past Lessons and Future Implications," in Sandlund, Schei, and Viken, op. cit. note 9; Allen, op. cit. note 9; Statement of Senator John Glenn, Senate Environment and Public Works Committee, Subcommittee on Drinking Water, Fisheries and Wildlife, Washington, DC, 19 September 1996; John Tibbetts, "Malaria Dreams: El Niño and Infectious Diseases," *Coastal Heritage,* summer 1996.

11. Fleet size and trade volume from U.N. Conference on Trade and Development (UNCTAD), *Review of Maritime Transport 1997* (New York: United Nations, 1997); 80 percent from Committee on Ships' Ballast Operations, op. cit. note 6.

12. Ballast water biota from James T. Carlton, "Invasions in the World's Seas: Six Centuries of Re-organizing Earth's Marine Life," in Sandlund, Schei, and Viken, op. cit. note 9, from Carlton and Geller, op. cit. note 9, and from Tirschwell, op. cit. note 4, quoting James Carlton.

13. Mark Williamson, *Biological Invasions* (London: Chapman and Hall, 1996); introduction from Carlton, op. cit. note 9; fishery collapse from Fred Pearce, "How the Soviet Seas Were Lost," *New Scientist*, 11 November 1995, and from Ivan Zrajevskij, "Black Sea Fisheries—A Case of Environmental Overload," *DHA News* (U.N. Department of Humanitarian Affairs), November/December 1995; annual fish catch from U.N. Food and Agriculture Organization (FAO), *FAO Fishery Statistics: Catches and Landings* (Rome:1995).

14. Pearce, op. cit. note 13; Neal Ascherson, *Black Sea* (New York: Hill and Wang, 1995); John Travis, "Invader Threatens Black, Azov Seas," *Science*, 26 November 1993.

15. Decline of Black Sea infestation from Ascherson, op. cit. note 14, and from Williamson, op. cit. note 13; movement out of the Black Sea from Ascherson, op. cit. note 14, from Zrajevskij, op. cit. note 13, and from "Baltic News," *Aliens* (Invasive Species Specialist Group, Species Survival Commission, IUCN), September 1995; Black Sea jellies on North American coast from Committee on Ships' Ballast Operations, op. cit. note 6.

16. GREENLines (Grassroots Environmental Effectiveness Network, a project of Defenders of Wildlife), 11 December 1997; oil development from Great Lakes Panel on Aquatic Nuisance Species, "Biological Invasions," flier (August 1996), and from Leann Ferry, "Preventing Nonindigenous Species Invasions in Prince William Sound, Alaska," *Aquatic Nuisance Species Digest*, August 1996.

17. Montpelier from K.V. Sykora, "History of the Impact of Man on the Distribution of Plant Species," in F. di Castri, A.J. Hansen, and M. Debussche, eds., *Biological Invasions in Europe and the Mediterranean Basin* (Dordrecht, the Netherlands: Kluwer Academic Publishers, 1990); cordgrass from Cory Dean, "Salt Marsh Interloper Alters a Coastline," *New York Times*, 9 March 1991; Chinese tallow tree from Bruce A. Stein and Stephanie R. Flack, eds., *America's Least Wanted: Alien Species Invasions of U.S. Ecosystems* (Arlington, VA: The Nature Conservancy, 1996), and from G. Jubinsky and L.C. Anderson, "The Invasive Potential of the Chinese Tallow-Tree (*Sapium sebiferum* Roxb.) in the Southeast," *Castanea*, vol. 61, no. 3 (1996).

18. Peter Jenkins, "Free Trade and Exotic Species Introductions," *Aliens* (Invasive Species Specialist Group, Species Survival

Commission, IUCN), March 1996, citing OTA, op. cit. note 9.

19. FAO, *State of the World's Forests 1997* (Rome: 1997).

20. Conference from Larry Lack, "Pest Plagues from Imported Wood," *Earth Island Journal*, fall 1996.

21. Import regulations from Faith Campbell, exotic species expert, Western Ancient Forests Campaign, e-mail to author, May 1998, and from Oregon Natural Resources Council et al., "Court Rules Log Import Regs Violate Law," press release (Portland, OR: 4 March 1997); Forest Service evaluation quoted in Faith Thompson Campbell and Scott E. Schlarbaum, *Fading Forests: North American Trees and the Threat of Exotic Pests* (New York: Natural Resources Defense Council, 1994).

22. Campbell and Schlarbaum, op. cit. note 21.

23. Jon R. Luoma, "Catfight," *Audubon*, July/August 1997; Seth Mydans, "The Stray Cats of Australia: 9 Lives Seen as 9 Too Many," *New York Times*, 28 January 1997; Alan Burdick, "Attack of the Aliens: Florida Tangles with Invasive Species," *New York Times*, 6 July 1996; Elizabeth Royte, "On the Brink: Hawaii's Vanishing Species," *National Geographic*, September 1995.

24. Stein and Flack, op. cit. note 17; Burdick, op. cit. note 23.

25. Peter B. Moyle, "Effects of Invading Species on Freshwater and Estuarine Ecosystems," in Sandlund, Schei, and Viken, op. cit. note 9.

26. Breeding from H.P.C. Shetty, M.C. Nandeesha, and A.G. Jhingran, "Impact of Exotic Aquatic Species in Indian Waters," from K.J. Ang, R. Gopinath, and T.E. Chua, "The Status of Introduced Fish Species in Malaysia," and from Twesukdi Piyakarnchana, "Exotic Aquatic Species in Thailand," all in Sena S. de Silva, ed., *Exotic Aquatic Organisms in Asia*, Asian Fisheries Society (AFS) Special Publication 3 (Manila: AFS, 1989), and from Walter R. Courtenay, Jr., "Biological Pollution Through Fish Introductions," in Bill N. McKnight, ed., *Biological Pollution: The Control and Impact of Invasive Exotic Species* (Indianapolis, IN: Indiana Academy of Science, 1993); aquarium introductions into the United States from ibid.; Asian eel from Rick Gragg, "A Wary Georgia Prepares to Battle the Kudzu of the Animal Kingdom," *New York Times*, 8 December 1996.

27. William F. Laurance et al., "In Defense of the Epidemic Disease Hypothesis," *Conservation Biology*, August 1997.

28. Carlton, op. cit. note 9; Tara Patel, "'Metamorphosis' Paper Greeted with Derision," *New Scientist*, 10 February 1996; Marlise Simons, "A Delicate Pacific Seaweed Is Now a Monster of the Deep," *New York Times*, 18 August 1996; size of the invasion from "Scientists Seek Authority to Widen Study of

Using Bioremediation to Halt Algae," *International Environment Reporter*, 9 July 1997.

29. Edward L. Mills et al., "Exotic Species in the Hudson River Basin: A History of Invasions and Introductions," *Estuaries*, December 1996; Charles F. Boudouresque, "The Red Sea–Mediterranean Link: Unwanted Effects of Canals," in Sandlund, Schei, and Viken, op. cit. note 9; Bella S. Galil, "Biodiversity and Invasion—How Resilient Is the Levant Sea?" *Israel Environment Bulletin*, vol. 20, no. 1 (1997).

30. Carlton, op. cit. note 12.

31. Henrik Jörgensen, "Control of Alien Species in Denmark, Legislation and Practical Experience," in Sandlund, Schei, and Viken, op. cit. note 9; Moyle, op. cit. note 25.

32. Ship sizes from Wiley, op. cit. note 7; increases in ballast releases from Carlton, op. cit. note 9.

33. Syren advertisement reproduced in James T. Carlton, "*Blue Immigrants:* The Marine Biology of Maritime History," *Mystic Seaport Log*, vol. 44 (1992), p. 33; steamer crossing time from Edward Tenner, *Why Things Bite Back: Technology and the Revenge of Unintended Consequences* (New York: Alfred A. Knopf, 1996); railroads from Robert R. Stickney, *Aquaculture in the United States: A Historical Survey* (New York: John Wiley and Sons, 1996).

34. International Civil Aviation Organization, "Growth in Air Traffic to Continue: ICAO Releases Long-Term Forecasts," press release (Montreal: March 1997); *1991 Information Please Almanac* (Boston: Houghton Mifflin); *New York Times 1998 Almanac* (New York: Penguin, 1997).

35. Carlton, op. cit. note 33.

36. Sparser fouling communities from ibid., and from J.T. Carlton and J. Hodder, "Biogeography and Dispersal of Coastal Marine Organisms: Experimental Studies on a Replica of a 16th-Century Sailing Vessel," *Marine Biology*, vol. 121 (1995), p. 722; other shipboard pathways from Committee on Ships' Ballast Operations, op. cit. note 6; Asian strain of the gypsy moth from OTA, op. cit. note 9; Jane E. Brody, "Invader From Asia Increases Gypsy Moth Threat," *New York Times*, 30 May 1995.

37. OTA, op. cit. note 9.

38. UNCTAD, op. cit. note 11; UNCTAD, *Review of Maritime Transport 1993* (New York: United Nations, 1994).

39. Types of organisms moving in containers from OTA, op. cit. note 9; George B. Craig, Jr., "The Diaspora of the Asian Tiger Mosquito," in McKnight, op. cit. note 26; used tire trade from Leslie Lamarre, "Tapping the Tire Pile," *EPRI Journal*,

September/October 1995, and from Michael Blumenthal and John Serumgard, "Scrap Tire Markets: Still Rollin' Along," *Resource Recycling*, March 1996.

40. Dengue from Anne E. Platt, "Confronting Infectious Diseases," in Lester R. Brown et al., *State of the World 1996* (New York: W.W. Norton & Company, 1996); other viruses from Chester G. Moore and Carl J. Mitchell, "*Aedes albopictus* in the United States: Ten-Year Presence and Public Health Implications," *Emerging Infectious Diseases*, July 1997; possible involvement in epidemics from Craig, op. cit. note 39.

41. Craig, op. cit. note 39.

42. New sets of pathways is based on a suggestion by Peter Vitousek, a Stanford University ecologist and expert on bioinvasions, who is quoted in John Tibbets, "Exotic Species: The Aliens Have Landed," *Coastal Heritage*, Spring 1997.

43. Institute of Medicine (IOM), Board of International Health, *America's Vital Interest in Global Health* (Washington, DC: National Academy Press, 1997).

44. Pat Hanlon, *Global Airlines: Competition in a Transnational Industry* (Oxford: Butterworth-Heinemann, 1996), reviewed in *Future Survey*, May 1996; International Air Transport Association, "Air Traffic on Track for 6.6% Growth for 2001" and "522 Million Sched. International Passengers in Year 2000," press releases (Montreal: 17 September 1997 and 28 October 1996); "Two U.S. Malaria Cases Spark Concern," *Washington Post*, 17 September 1995.

45. Immigration (1994 figure) from Laurie Garrett, "The Return of Infectious Disease," *Foreign Affairs*, January/February 1996; refugees and internally displaced from Jennifer D. Mitchell, "Refugee Flows Drop Steeply," in Lester R. Brown, Michael Renner, and Christopher Flavin, *Vital Signs 1998* (New York: W.W. Norton & Company, 1998); IOM, op. cit. note 43.

46. Resurgence of older diseases from Paul Epstein, "The Threatened Plague," *People and the Planet*, vol. 6, no. 3 (1997).

47. Complex of factors from Anne E. Platt, *Infecting Ourselves: How Environmental and Social Disruptions Trigger Disease*, Worldwatch Paper 129 (Washington, DC: Worldwatch Institute, April 1996); cholera from Platt, op. cit. note 40; drug resistance from Epstein, op. cit. note 46, and from IOM, op. cit. note 43; lack of drugs from ibid.; global mortality from Anne E. Platt, "Infectious Diseases Return," in Lester R. Brown, Christopher Flavin, and Hal Kane, *Vital Signs 1996* (New York: W.W. Norton & Company, 1996).

48. Tara Patel, "Yellow for Danger," *New Scientist*, 10 May 1997.

49. Dengue hemorrhagic fever from Craig, op. cit. note 39, and

from D.J. Gubler and D.W. Trent, "Emergence of Epidemic Dengue/Dengue Hemorrhagic Fever as a Public Health Problem in the Americas," *Infectious Agents and Disease*, vol. 2, no. 6 (1993); infection rates from Platt, "Infectious Diseases Return," op. cit. note 47; AIDS-malaria synergism from Pallava Bagla, "Malaria Fighters Gather at Site of Early Victory," *Science*, 5 September 1997; malaria infection rates from IOM, op. cit. note 43, and from Platt, op. cit. note 40.

50. P.C.C. Garnham, "The Origins, Evolutionary Significance and Dispersal of Haemosporidian Parasites of Domestic Birds," in J.M. Cherrett and G.R. Sagar, eds., *Origins of Pest, Parasite, Disease and Weed Problems* (Oxford: Blackwell Scientific Publications, 1977).

CHAPTER 8. Economic Invasions

1. U.S. Congress, Office of Technology Assessment (OTA), *Harmful Nonindigenous Species in the United States* (Washington, DC: September 1993); Jeffrey A. McNeely, "Costs and Benefits of Alien Species," in Odd Terje Sandlund, Peter Johan Schei, and Åslaug Viken, eds., *Proceedings of the Norway/UN Conference on Alien Species, Trondheim, 1–5 July 1996* (Trondheim: Directorate for Nature Management and Norwegian Institute for Nature Research, 1996); Peter Jenkins, Biopolicy Consulting, Placitas, NM, discussion with author, May 1998.

2. OTA, op. cit. note 1; Jenkins, op. cit. note 1.

3. Value of crops and share lost to pests from David Pimentel et al., "Economic and Environmental Benefits of Biodiversity," *BioScience*, December 1997, and from George N. Agrios, *Plant Pathology*, 4th ed. (San Diego, CA: Academic Press, 1997); pesticide expenditures from "World Agchem Market Recovery Continues," *Agrow*, 11 July 1997; percentage of exotic pests from Pimentel et al., op. cit. this note (20–70 percent), and from OTA, op. cit. note 1 (40–90 percent).

4. U.S. Department of Agriculture, Animal and Plant Health Inspection Service (APHIS) Plant Protection and Quarantine, "Medfly Fact Sheet," <http//www.aphis.usda.gov/oa/medflyfs. html>, viewed 16 December 1997; cotton bollworm from "Insecticide Resistance Management: Consider the Alternative," *Resistant Pest Management*, summer 1996, and from U.N. Food and Agriculture Organization (FAO), *FAO Production Yearbook 1995 and 1996* (Rome: 1996 and 1997).

5. Effects of the Russian wheat aphid are mentioned in J.R.S. Fincham, "The Killing Fields," *Nature*, 28 September 1995.

6. Monsanto's transgenic crops from "What's Coming To Market?" and "1997 Global Acreage of Monsanto's Transgenic Crops," both in *Gene Exchange* (agricultural biotechnology newsletter of the Union of Concerned Scientists, fall 1997); 1998 projection from Monsanto Company, *1997 Annual Report* (St. Louis, MO: 1997).

7. Gary Gardner, *Shrinking Fields: Cropland Loss in a World of Eight Billion*, Worldwatch Paper 131 (Washington, DC: Worldwatch Institute, July 1996); herbicide-tolerant crops and no-till agriculture from Monsanto Company, *Monsanto 1997 Report on Sustainable Development* (St. Louis, MO: March 1998).

8. Ricardo Carrere and Larry Lohmann, *Pulping the South: Industrial Tree Plantations and the World Paper Economy* (London: Zed Books, 1996).

9. Seth Mydans, "Thai Shrimp Farmers Facing Ecologists' Fury," *New York Times*, 28 April 1996; Will Nixon, "Rainforest Shrimp," *Mother Jones*, March/April 1996; Ehsan Masood, "Aquaculture: A Solution, or Source of New Problems?" *Nature*, 13 March 1997; Donald V. Lightner et al., "Geographic Dispersion of the Viruses IHHN, MBV and HPV as a Consequence of Transfers and Introductions of Penaeid Shrimp to New Regions for Aquaculture Purposes," in Aaron Rosenfield and Roger Mann, eds., *Dispersal of Living Organisms into Aquatic Ecosystems* (College Park, MD: Maryland Sea Grant College, 1992); Beena Pandey and Sachin Chaturvedi, "Prospects for Aquaculture in India," *Biotechnology and Development Monitor*, December 1994; "Another Shrimp 'Gold Rush' Goes Bust," *Ceres*, November–December 1995; Sam Howe Verhovek, "Virus Imperils Texas Shrimp Farms," *New York Times*, 14 June 1995.

10. Ivan Zrajevskij, "Black Sea Fisheries—A Case of Environmental Overload," *DHA News*, November/December 1995; Fred Pearce, "How the Soviet Seas Were Lost," *New Scientist*, 11 November 1995; John Travis, "Invader Threatens Black, Azov Seas," *Science*, 26 November 1993. Losses to the jelly totaled $250 million by 1993. At $30 million a year for subsequent years, the total reached $370 million by the end of 1997.

11. Pest losses in North American forests from David Pimentel et al., "Environmental and Economic Costs of Introduced Nonindigenous Species in the United States," draft (Ithaca, NY: Cornell University, 1998); gypsy moth from OTA, op. cit. note 1; current expenditures on gypsy moth from Richard Fowler, Office of Forest Pest Management, U.S. Forest Service, discussion with author, April 1998.

12. Brown tree snake from Mark Jaffe, *And No Birds Sing: The Story of an Ecological Disaster in a Tropical Paradise* (New York: Simon and Schuster, 1994), and from Thomas H. Fritts and Gordon H. Rodda, "Trouble in Paradise: The Brown Tree Snake in the Western Pacific," *Aquatic Nuisance Species Digest*, March 1996; Formosan termite from Alan Holt, "An Alliance of Biodiversity, Agriculture, Health, and Business Interests for Improved Alien Species Management in Hawaii," in Sandlund, Schei, and Viken, op. cit. note 1, and from "Introduction: Exotic Pests," and Michael K. Rust et al., "Ravenous Formosan Subterranean Termites Persist in California," both in *California Agriculture*, March–April 1998.

13. John Cairns, Jr., and Joseph R. Bidwell, "Discontinuities in Technological and Natural Systems Caused by Exotic Species," *Biodiversity and Conservation*, vol. 5 (1996), p. 1,088.

14. Blocking of pipes from Michael L. Ludyanskiy, Derek McDonald, and David MacNeill, "Impact of the Zebra Mussel, a Bivalve Invader," *BioScience*, September 1993; mussel removal from Charles R. O'Neill, Jr., "Economic Impact of Zebra Mussels—Results of the 1995 National Zebra Mussel Information Clearinghouse Study," *Great Lakes Research Review*, April 1997; raw water user costs from "Zebra Mussels Cost Great Lakes Water Users an Estimated $120 Million," *ANSUpdate* (winter 1996), insert in *Aquatic Nuisance Species Digest*, March 1996; power plants from O'Neill, op. cit. this note; permanent costs from Cairns and Bidwell, op. cit. note 13.

15. Cumulative total costs from OTA, op. cit. note 1 ($3.1 billion in 1991 dollars over a decade); Ludyanskiy, McDonald, and MacNeill, op. cit. note 14, quote an unpublished 1989 Fish and Wildlife Service estimate of $5 billion.

16. "Southern African Environmental Issues No. 11: Water Hyacinth," factsheet (Harare, Zimbabwe: Communicating the Environment Programme, 1995); "Beating the Dam Busters," *African Farming*, March/April 1997.

17. Lars W.J. Anderson, "Eradicating California's Hydrilla," *Aquatic Nuisance Species Digest*, March 1996; Jeffery Schardt, "Nonindigenous Aquatic Weeds: A National Problem," *Aquatic Nuisance Species Digest*, July 1995.

18. McNeely, op. cit. note 1.

19. Peter Friederici, "The Alien Saltcedar," *American Forests*, January/February 1995; William Wiesenborn, "*Tamarix ramosissima, T. chinensis, T. parviflora:* Tamarisk," in John M. Randall and Janet Marinelli, eds., *Invasive Plants: Weeds of the Global Garden* (New York: Brooklyn Botanic Garden, 1996).

20. Edward Tenner, *Why Things Bite Back: Technology and the*

Revenge of Unintended Consequences (New York: Alfred A. Knopf, 1996).

21. Introduction of melaleuca from George F. Meskimen, "A Silvical Study of the Melaleuca Tree in South Florida," Ph.D. dissertation (Gainesville: University of Florida, 1962); fire adaptations from Ronald L. Myers, "Site Susceptibility to Invasion by the Exotic Tree *Melaleuca quinquenervia* in Southern Florida," *Journal of Applied Ecology*, vol. 20 (1983), pp. 645–58, from Julia F. Morton, "The Cajeput Tree—A Boon and an Affliction," *Economic Botany*, vol. 20 (1966), pp. 31–39, and from Eric Morgenthaler, "What's Florida to Do With an Explosion of Melaleuca Trees?" *Wall Street Journal*, 8 February 1993.

22. Fire at 58th Street from John D. Flowers, II, "Subtropical Fire Suppression in *Melaleuca quinquenervia*," in Ted D. Center et al., eds., *Proceedings of the Symposium on Exotic Pest Plants* (Washington, DC: U.S. Department of the Interior, National Park Service, 1991); distribution of melaleuca from Amy Ferriter et al., "Management in Water Management Districts," in Daniel Simberloff, Don C. Schmitz, and Tom C. Brown, eds., *Strangers in Paradise: Impact and Management of Non-indigenous Species in Florida* (Washington, DC: Island Press, 1997).

23. Losses in Peru from Paul Epstein, "The Threatened Plague," *People and the Planet*, vol. 6, no. 3 (1997); water expenditures from Laurie Garrett, *The Coming Plague: Newly Emerging Diseases in a World Out of Balance* (New York: Farrar, Straus and Giroux, 1994), citing Pan American Health Organization (PAHO) figures.

24. PAHO and the yellow fever mosquito from Ralph T. Bryan, "Alien Species and Emerging Infectious Diseases: Past Lessons and Future Implications," in Sandlund, Schei, and Viken, op. cit. note 1; Anne E. Platt, "Infectious Diseases Return," in Lester R. Brown, Christopher Flavin, and Hal Kane, *Vital Signs 1996* (New York: W.W. Norton & Company, 1996).

25. D.J. Gubler and D.W. Trent, "Emergence of Epidemic Dengue/Dengue Hemorrhagic Fever as a Public Health Problem in the Americas," *Infectious Agents and Disease*, vol. 2, no. 6 (1993); Gary Taubes, "A Mosquito Bites Back," *New York Times Magazine*, 24 August 1997; breeding sites of the Asian tiger mosquito from George B. Craig, Jr., "The Diaspora of the Asian Tiger Mosquito," in Bill N. McKnight, ed., *Biological Pollution: The Control and Impact of Invasive Exotic Species* (Indianapolis, IN: Indiana Academy of Science, 1993).

26. Cost of Lee County mosquito control from Taubes, op. cit. note

25.

27. C.E. Hughes, "Protocols for Plant Introductions with Particular Reference to Forestry: Changing Perspectives on Risks to Biodiversity and Economic Development," in Charles H. Stirton, chair, *Weeds in a Changing World*, Symposium Proceedings 64 (Farnham, U.K.: British Crop Protection Council, 1996).

28. Mydans, op. cit. note 9; Anne Platt McGinn, "Blue Revolution: The Promises and Pitfalls of Fish Farming," *World Watch*, March/April 1998; Gary Cohen, "Aquaculture Floods Indian Villages," *Multinational Monitor*, July/August 1995.

29. Phyllis Windle, letter to the editor, *Issues in Science and Technology*, fall 1997; Faith Campbell, exotic species expert, Western Ancient Forests Campaign, e-mail to author, May 1998.

30. V.M.G. Nair, "Oak Wilt: An Internationally Dangerous Tree Disease," in S.P. Raychaudhuri and Karl Maramorosch, eds., *Forest Trees and Palms: Diseases and Control* (Lebanon, NH: Science Publishers, 1996); Agrios, op. cit. note 3; Karl Maramorosch, "The Cadang-Cadang Disease of Palms," in Raychaudhuri and Maramorosch, op. cit. this note; FAO, *Trade Yearbook 1995* (Rome: 1995); Wade Davis, "The Rubber Industry's Biological Nightmare," *Fortune*, 4 August 1997.

31. U.S. federal inspection and quarantine expenses from U.S. General Accounting Office, *Agricultural Inspection: Improvements Needed to Minimize Threat of Foreign Pests and Diseases* (Washington, DC: May 1997) (APHIS expenditures on agricultural quarantine and inspection came to $151.9 million in fiscal year 1996); Alan Burdick, "Attack of the Aliens: Florida Tangles with Invasive Species," *New York Times*, 6 June 1995; Daniel Simberloff, Don C. Schmitz, and Tom C. Brown, "Why We Should Care and What We Should Do," in Simberloff, Schmitz, and Brown, op. cit. note 22; Jeffery Schardt, "Nonindigenous Aquatic Weeds: A National Problem," *Aquatic Nuisance Species Digest*, July 1995.

32. OTA, op. cit. note 1.

33. Molly O'Meara, "Global Temperature Reaches Record High," in Lester R. Brown, Michael Renner, and Christopher Flavin, *Vital Signs 1998* (New York: W.W. Norton & Company, 1998); Robert T. Watson et al., "Technical Summary: Impacts, Adaptations, and Mitigation Options," in Robert T. Watson et al., eds., *Climate Change 1995: Impacts, Adaptations and Mitigation of Climate Change: Scientific-Technical Analyses: Contribution of Working Group II to the Second Assessment Report of the Intergovernmental Panel on Climate Change* (New York:

Cambridge University Press, 1996).

34. Barbara Allen-Diaz et al., "Rangelands in a Changing Climate: Impacts, Adaptations, and Mitigation," in Watson et al., op. cit. note 33; Bob Holmes, "Unwelcome Guests," *New Scientist*, 18 April 1998; Chris Bright, "Tracking the Ecology of Climate Change," in Lester R. Brown et al., *State of the World 1997* (New York: W.W. Norton & Company, 1997).

35. Holmes, op. cit. note 34; Bright, op. cit. note 34.

36. Werner A. Kurz et al., "Global Climate Change: Disturbance Regimes and Biospheric Feedbacks of Temperate and Boreal Forests," in George M. Woodwell and Fred T. MacKenzie, eds., *Biotic Feedbacks in the Global Climatic System: Will the Warming Feed the Warming?* (New York: Oxford University Press, 1995); Bright, op. cit. note 34.

37. Michael J. Samways, "Managing Insect Invasions by Watching Other Countries," in Sandlund, Schei, and Viken, op. cit. note 1.

CHAPTER 9. Toward an Ecologically Literate Society

1. Joachim Wolschke-Bulmahn, "The Mania for Native Plants in Nazi Germany," in Mark Dion and Alexis Rockman, eds., *Concrete Jungle* (New York: Juno Books, 1996).

2. Colorado potato beetle from Edward Tenner, *Why Things Bite Back: Technology and the Revenge of Unintended Consequences* (New York: Alfred A. Knopf, 1996); Pascal Fletcher, "Cuba Sees US Hand Behind Insect Plague," *Financial Times*, 12 May 1997.

3. Hilary F. French, "Environmental Treaties Grow in Number," in Lester R. Brown, Nicholas Lenssen and Hal Kane, *Vital Signs 1995* (New York: W.W. Norton & Company, 1995); Lynton Keith Caldwell, *International Environmental Policy*, 2nd rev. ed. (Durham, NC: Duke University Press, 1990).

4. Caldwell, op. cit. note 3.

5. Christopher Flavin and Seth Dunn, *Rising Sun, Gathering Winds: Policies to Stabilize the Climate and Strengthen Economies*, Worldwatch Paper 138 (Washington, DC: Worldwatch Institute, November 1997); Table 9–1 from Lyle Glowka and Cyrille de Klemm, "International Instruments, Processes, Organizations and Non-Indigenous Species Introductions: Is a Protocol to the Convention on Biological Diversity Necessary?" in Odd Terje Sandlund, Peter Johan Schei, and Åslaug Viken, eds., *Proceedings of the Norway/UN Conference on Alien Species, Trondheim, 1–5 July 1996* (Trondheim: Directorate

for Nature Management and Norwegian Institute for Nature Research, 1996). Glowka and de Klemm list 22 agreements; I added the International Plant Protection Convention (IPPC).

6. Glowka and de Klemm, op. cit. note 5.

7. Idea of expanding the IPPC from Faith Campbell, presentation at the Symposium on the Control and Impact of Invasive Exotic Species, Indianapolis, IN, October 1991.

8. Glowka and de Klemm, op. cit. note 5; Faith Campbell, exotic species expert for the Western Ancient Forests Campaign, and Hon. Jim Jontz, U.S. House of Representatives, "Memo on the Present Status of the IPPC and Upcoming 'Review' of the SPS Agreement" (Washington, DC: 1 December 1997). The World Trade Organization sanitary agreement is the 1994 World Trade Organization Agreement on the Application of Sanitary and Phytosanitary Standards.

9. Current status of the convention from <http://www.un.org/law>; analysis of Article 196 from U.S. Congress, Office of Technology Assessment (OTA), *Harmful Nonindigenous Species in the United States* (Washington, DC: September 1993).

10. Current status of the convention from <http://www.unccd.ch>, under "Other Web Sites"; International Union for Conservation of Nature and Natural Resources (IUCN), *Biological Diversity Convention Draft (June 30, 1989)* (Bonn: IUCN Environmental Law Centre, 1989); Peter T. Jenkins, "Free Trade and Exotic Species Introductions," *Conservation Biology*, February 1996; Peter T. Jenkins, "Harmful Exotics in the United States," in William J. Snape III, ed., *Biodiversity and the Law* (Washington, DC: Island Press, 1996).

11. Expert exotics panel from Jenkins, "Harmful Exotics in the United States," op. cit. note 10.

12. Scott Hajost and Curtis Fish, "Biodiversity Conservation and International Instruments," in Snape, op. cit. note 10; Glowka and de Klemm, op. cit. note 5.

13. International Council for Local Environmental Initiatives, "Local Government Implementation of Agenda 21," <http://www.iclei.org>, viewed April 1997; Hajost and Fish, op. cit. note 12; Glowka and de Klemm, op. cit. note 5.

14. Harold A. Mooney, "The SCOPE Initiative: The Background and Plans for a Global Strategy on Invasive Species," in Sandlund, Schei, and Viken, op. cit. note 5; Michael N. Clout and Sarah J. Lowe, "Reducing the Impacts of Invasive Species on Global Biodiversity: The Role of the IUCN Invasive Species Specialist Group," in ibid.; IUCN Species Survival Commission, *The IUCN Position Statement on the Translocation of Living Organisms* (Gland, Switzerland: IUCN, September

1987). The Invasive Species Specialist Group (ISSG) draft guidelines are available online at <http://www.iucn.org/themes/ssc/memonly/invguid.htm>.

15. Jarle Mork, "Control Measures Regarding Marine Invasions. The ICES Code of Practice on the Introductions and Transfers of Marine Organisms 1994," in Sandlund, Schei, and Viken, op. cit. note 5; U.N. Food and Agriculture Organization (FAO), *Code of Conduct for Responsible Fisheries* (Rome: 1995), articles 9.2.3, 9.3.1, and 9.3.2, p. 24; FAO, *Precautionary Approach to Capture Fisheries and Species Introductions*, FAO Technical Guidelines for Responsible Fisheries 2 (Rome: 1996) (the International Council for the Exploration of the Sea code is included as Appendix A).

16. Untested tropical trees from Rebecca P. Butterfield, "Promoting Biodiversity: Advances in Evaluating Native Species for Reforestation," *Forest Ecology and Management*, vol. 75 (1995), pp. 111–21; native North American pollinators from Stephen L. Buchman and Gary Paul Nabhan, *The Forgotten Pollinators* (Washington, DC: Island Press/Shearwater Books, 1996), from David Pimentel et al., "Economic and Environmental Benefits of Biodiversity," *BioScience*, December 1997, and from Sue Hubbell, "Trouble with Honeybees," *Natural History*, May 1997.

17. The SCOPE plan envisions an information "clearing house mechanism"; see Mooney, op. cit. note 14.

18. Readers interested in exploring this environment can find an entrée through the following addresses (current at the time of writing): the U.S. Department of Agriculture's Animal and Plant Health Inspection Service noxious weeds home page, <http://www.aphis.usda.gov/oa/weeds/weedhome.html>; the U.S. Geological Survey nonindigenous aquatic species home-page, <http://nas.nfrcg.gov/nas.html>; the home page of the Centre for Agriculture and Biosciences, International, <http://pest.cabweb.org>; the ProMed newservice on emerging diseases, <http://www.healthnet.org/programs/promed.html>; or the ambitiously named "World Species List—Animals Plants Microbes," <http://www.envirolink.org/species>.

19. ISSG project from Clout and Lowe, op. cit. note 14. Some databases already on-line could serve as useful models, such as the databases available through the Centre for Agriculture and Biosciences, International, op. cit. note 18, and Australian National University Virus Databases On-Line, <http://life.anu.edu.au/viruses/welcome.htm>.

20. Both the acts themselves and a set of interpretative information sheets are available on the New Zealand Ministry for the

Environment home page, <http://www.mfe.govt.nz/>. See also Curtis C. Daehler and Doria R. Gordon, "To Introduce or Not To Introduce: Trade-Offs of Non-indigenous Organisms," *TREE* (Trends in Ecology and Evolution), November 1997.

21. Michigan from John Niyo, "Alien Species Run Amok," *Ann Arbor News*, 13 August 1995; on the general problem, see, for example, Ted Williams, "Invasion of the Aliens," *Audubon*, September/October 1994, and Douglas B. Houston and Edward G. Schreiner, "Alien Species in National Parks: Drawing lines in Space and Time," *Conservation Biology*, February 1995.

22. Adherence to protocols from Jenkins, "Harmful Exotics in the United States," op. cit. note 10.

23. Argentine ant from Luc Passera, "Characteristics of Tramp Species," in David F. Williams, ed., *Exotic Ants: Biology, Impact, and Control of Introduced Species* (Boulder, CO: Westview Press, 1994), from J.D. Majer, "Spread of Argentine Ants *(Linepithema humile)*, with Special Reference to Western Australia," in ibid., and from C. Mlot, "Invasive Argentine Ant Is No Picnic," *Science News*, 23 August 1997; green crab from Andrew N. Cohen, "Have Claw, Will Travel," *Aquatic Nuisance Species Digest*, August 1997).

24. *National Invasive Species Act of 1996*, Public Law 104-332, is a set of amendments to the *Nonindigenous Aquatic Nuisance Species Prevention and Control Act of 1990*, Public Law 101-646; alternative ballast water techniques from Committee on Ships' Ballast Operations, National Research Council, *Stemming the Tide: Controlling Introductions of Nonindigenous Species by Ships' Ballast Water* (Washington, DC: National Academy Press, 1996).

25. George B. Craig, Jr., "The Diaspora of the Asian Tiger Mosquito," in Bill N. McKnight, ed., *Biological Pollution: The Control and Impact of Invasive Exotic Species* (Indianapolis, IN: Indiana Academy of Science, 1993); Chester G. Moore and Carl J. Mitchell, "*Aedes albopictus* in the United States: Ten-Year Presence and Public Health Implications," *Emerging Infectious Diseases*, July 1997.

26. Jane H. Bock and Carl E. Bock, "The Challenges of Grassland Conservation," in Anthony Joern and Kathleen H. Keeler, eds., *The Changing Prairie: North American Grasslands* (New York: Oxford University Press, 1995); scarcity of native grass seed from Maureen Kuwano Hinkle, Director, Agricultural Policy, National Audubon Society, note to author, April 1998.

27. Francis M. Harty, "How Illinois Kicked the Exotic Habit," in McKnight, op cit. note 25.

28. Noel Vietmeyer, *Lost Crops of Africa, vol. 1: Grains* (Washington,

DC: National Academy Press, 1996).

29. "Concerned about Forests, Suburbanites Block Prairie Project," *New York Times*, 7 November 1996; Lou Cannon, "California Fish Kill Mounts after Poisoning by State," *Washington Post*, 17 October 1997.

30. Stanley A. Temple, "The Nasty Necessity: Eradicating Exotics," *Conservation Biology*, June 1990.

31. George Molnar et al., "Management of *Melaleuca quinquenervia* within East Everglades Wetlands," in Ted D. Center et al., eds., *Proceedings of the Symposium on Exotic Pest Plants* (Washington, DC: U.S. Department of Interior, National Park Service, 1991).

32. South Africa Department of Water Affairs and Forestry, *The Working for Water Programme Annual Report 1995/96* (Johannesburg: 1996).

33. New Zealand from Dick Veitch, "Preventing Rat Resurgence," *Aliens* (newsletter of the ISSG, Species Survival Commission, IUCN), September 1996; Mark Day and Jennifer Daltry, "Rat Eradication to Conserve the Antiguan Racer," *Aliens*, March 1996; island rat "naiveté" from Ian McFadden, "Rodent Eradication from New Zealand Islands," *Aliens*, March 1995.

34. Bruce Coblentz, "Judas Goats in the Tropics," *Aliens* (newsletter of the ISSG, Species Survival Commission, IUCN), March 1995.

35. Ibid.

36. Micronesian rat story from Christopher Lever, *Naturalized Animals: The Ecology of Successfully Introduced Species* (New York: Academic Press, 1995), reviewed in *Aliens* (newsletter of the ISSG, Species Survival Commission, IUCN), September 1995.

37. Ibid.; flatworm introduction from Lucius G. Eldredge and Barry D. Smith, "Triclad Flatworm Tours the Pacific," *Aliens* (newsletter of the ISSG, Species Survival Commission, IUCN), September 1995, and from R. Muniappan, "Use of the Planarian, *Platydemus manokwari*, and other Natural Enemies to Control the Giant African Snail," in Jan Bay-Petersen, Osamu Mochida, and Keizi Kiritani, eds., *The Use of Natural Enemies to Control Agricultural Pests* (Taipei: Food and Fertilizer Technology Center, 1990).

38. Quote from Henry R. Rupp, "Adverse Assessments of *Gambusia affinis*: An Alternative View for Mosquito Control Practitioners," *Journal of the American Mosquito Control Association*, vol. 12, no. 2 (1996); extinctions from Walter R. Courtenay, Jr., "Biological Pollution Through Fish Introductions," in McKnight, op. cit. note 25.

39. Ideal biocontrol agent from Paul Debach and David Rosen,

Biological Control by Natural Enemies, 2nd ed. (Cambridge, U.K.: Cambridge University Press, 1991); Hawaiian extinctions from Daniel S. Simberloff, "Community Effects of Introduced Species," in Matthew H. Nitecki, ed., *Biotic Crises in Ecological and Evolutionary Time* (New York: Academic Press, 1981), and from Francis G. Howarth, Gordon Nishida, and Adam Asquith, "Insects of Hawaii," in Edward T. LaRoe et al., eds., *Our Living Resources: A Report to the Nation on the Distribution, Abundance, and Health of U.S. Plants, Animals, and Ecosystems* (Washington, DC: U.S. Department of the Interior, National Biological Service, 1995).

40. Michael B. Usher, "Biological Invasions Into Tropical Nature Reserves," in P.S. Ramakrishnan, ed., *Ecology of Biological Invasions in the Tropics*, Proceedings of an International Workshop Held at Nainital, India (New Delhi: International Scientific Publications, 1989); Warren Herb Wagner, Jr., "Problems with Biotic Invasives: A Biologist's Viewpoint," in McKnight, op. cit. note 25.

41. Joby Warrick, "A Mite Strikes Mighty Blow Against World Hunger," *Washington Post*, 19 May 1997; J.S. Yaninek, A. Onzo, and J.B. Ojo, "Continent-Wide Releases of Neotropical Phytoseiids against the Exotic Cassava Green Mite in Africa," *IITA Research* (International Institute for Tropical Agriculture), March 1994.

42. Introduction of the Brazilian rabbit virus from George Laycock, *The Alien Animals* (Garden City, NY: Natural History Press, 1966); recent rabbit damage and escape of the other virus (rabbit hemorrhagic disease) from Jason Alexandra, "The Great Escape," *Habitat Australia*, April 1996, from Dan Drollette, "Australia Fends Off Critic of Plan to Eradicate Rabbits," *Science*, 12 April 1996, and from Ian Anderson, "Killer Rabbit Virus on the Loose," *New Scientist*, 21 October 1995.

43. Ecological recovery from Dan Drollette, "Wide Use of Rabbit Virus Is Good News for Native Species," *Science*, 10 January 1997; arid-land flea from Roger P. Pech, "Managing Alien Species: The Australian Experience," in Sandlund, Schei, and Viken, op. cit. note 5; cane toad from "Anti-Toad Virus to Be Introduced into Australia," *Pesticides News*, March 1996, and from Bryony Bennett, "Preparing for Battle with *Bufo marinus*," *Ecos*, spring 1996; general reluctance to use microbes from Peter Harris, "Environmental Impact of Introduced Biological Control Agents," in Manfred Mackauer, Lester E. Ehler, and Jens Roland, eds., *Critical Issues in Biological Control* (Andover, U.K.: Intercept, 1990).

44. For biocontrol success rates, see the surveys cited in OTA, op. cit. note 9, and John J. Drea, "Classical Biological Control—An Endangered Discipline?" in McKnight, op. cit. note 25; George Oduor, "Biological Pest Control and Invasives," in Sandlund, Schei, and Viken, op. cit. note 5.

45. Engineered viruses from Roger P. Pech, "Managing Alien Species: The Australian Experience," in Sandlund, Schei, and Viken, op. cit. note 5, and from Ian Anderson, "Alarm Greets Contraceptive Virus," *New Scientist*, 26 April 1997; sterile male releases in the medfly from Michael Ashburner, "Medfly Transformed—Official!" *Science*, 22 December 1995; mosquitoes from Stephen Young, "Mosquitoes That Kill Malaria," *New Scientist*, 5 August 1995, from Gautam Naik, "Turning Mosquitoes into Malaria Fighters," *Wall Street Journal*, 17 June 1997, and from Paul Recer, "Altered Virus Injected into Mosquitoes Blocks Dengue Fever," *Washington Post*, 10 May 1996.

46. Chris Viney, "Pest Control in the Deep," *Ecos*, spring 1996; Lisa Jones, "Sexy Weapon Thwarts Bugs," *High Country News*, 14 November 1994; Doug McInnis, "Birth Control Might Keep Mustangs from Outgrowing Range," *New York Times*, 26 March 1996.

47. Stephen R. Potter, *Commoners, Tribute, and Chiefs: The Development of Algonquian Culture in the Potomac Valley* (Charlottesville, VA: University of Virginia Press, 1993).

Index

ABOUT THE AUTHOR

CHRIS BRIGHT is a research associate at the Worldwatch Institute and senior editor of the Institute's magazine, *WorldWatch*. His research focuses on biodiversity issues. In addition to his work on the magazine, he has contributed chapters to the Institute's annual publication, *State of the World*, on bioinvasions, the ecological effects of climate change, and (with John Tuxill) global vertebrate decline. Before joining the Institute in 1994, Mr. Bright was as an editor at the *American Gardener* magazine, where his main concern was in finding ways to increase reader interest in environmental issues. His own articles for that magazine covered such topics as the natural history of native American trees, genetic engineering of fruit and vegetable crops, the effects of peat harvesting on bog ecosystems, and gardening for wildlife. Before becoming an editor, Mr. Bright wrote freelance for general circulation magazines on a broad range of environmental issues, from whaling and forestry to the pesticide trade. Mr. Bright's academic background is in Old English and Latin literature. He lives outside Washington, D.C., on land formerly occupied by the Tauxenent people, where he cultivates an interest in Zen Buddhism and a collection of exotic conifers.